Research Series on the Chinese Dream and China's Development Path

Drawing on a large body of empirical studies done over the last two decades, this Series provides its readers with in-depth analyses of the past and present and forecasts for the future course of China's development. It contains the latest research results made by members of the Chinese Academy of Social Sciences. This series is an invaluable companion to every researcher who is trying to gain a deeper understanding of the development model, path and experience unique to China. Thanks to the adoption of Socialism with Chinese characteristics, and the implementation of comprehensive reform and opening-up, China has made tremendous achievements in areas such as political reform, economic development, and social construction, and is making great strides towards the realization of the Chinese dream of national rejuvenation. In addition to presenting a detailed account of many of these achievements, the authors also discuss what lessons other countries can learn from China's experience.

More information about this series at http://www.springer.com/series/13571

Jingyuan Li • Tongjin Yang
Editors

China's Eco-city Construction

SSAP

Springer

Editors
Jingyuan Li
Center for Culture Study
Chinese Academy of Social Sciences
Beijing, China

Tongjin Yang
School of Public Policy and Management
Guangxi University
Nanning, China

ISSN 2363-6866 ISSN 2363-6874 (electronic)
Research Series on the Chinese Dream and China's Development Path
ISBN 978-3-662-48152-3 ISBN 978-3-662-48153-0 (eBook)
DOI 10.1007/978-3-662-48153-0

Library of Congress Control Number: 2015947766

Springer Heidelberg New York Dordrecht London

Printed on acid-free paper

Springer-Verlag GmbH Berlin Heidelberg is part of Springer Science+Business Media (www.springer.com)

Series Preface

Since China's reform and opening began in 1978, the country has come a long way on the path of Socialism with Chinese characteristics, under the leadership of the Communist Party of China. Over 30 years of reform efforts and sustained spectacular economic growth have turned China into the world's second-largest economy and wrought many profound changes in the Chinese society. These historically significant developments have been garnering increasing attention from scholars, governments and the general public alike around the world since the 1990s, when the newest wave of China Studies began to gather steam. Some of the hottest topics have included the so-called China miracle, Chinese phenomenon, Chinese experience, Chinese path and the Chinese model. Homegrown researchers have soon followed suit. Already hugely productive, this vibrant field is putting out a large number of books each year, with Social Sciences Academic Press alone having published hundreds of titles on a wide range of subjects.

Because most of these books have been written and published in Chinese, however, readership has been limited outside China – even among many who study China – for whom English is still the lingua franca. This language barrier has been an impediment to efforts by academia, business communities and policymakers in other countries to form a thorough understanding of contemporary China, of what is distinct about China's past and present may mean not only for her future but also for the future of the world. The need to remove such an impediment is both real and urgent, and the *Research Series on the Chinese Dream and China's Development Path* is my answer to the call.

This series features some of the most notable achievements from the last 20 years by scholars in China in a variety of research topics related to reform and opening. They include both theoretical explorations and empirical studies and cover economy, society, politics, law, culture and ecology, the six areas in which reform and opening policies have had the deepest impact and farthest-reaching consequences for the country. Authors for the series have also tried to articulate their visions of the "Chinese Dream" and how the country can realize it in these fields and beyond.

All of the editors and authors for the *Research Series on the Chinese Dream and China's Development Path* are both longtime students of reform and opening and recognized authorities in their respective academic fields. Their credentials and expertise lend credibility to these books, each having been subject to a rigorous peer review process for inclusion in the series. As part of the Reform and Development Program under the State Administration of Press, Publication, Radio, Film and Television of the People's Republic of China, the series is published by Springer, a Germany-based academic publisher of international repute, and distributed overseas. I am confident that it will help fill a lacuna in studies of China in the era of reform and opening.

Shouguang Xie

Series Acknowledgements

After a relatively short gestation period, the *Research Series on the Chinese Dream and China's Development Path* has started to bear fruits. We have, first and foremost, the book's authors and editors to thank for making this possible. And it was the hard work by many people at Social Sciences Academic Press and Springer, the two collaborating publishers, that made it a reality. We are deeply grateful to all of them.

Mr. Shouguang Xie, president of Social Sciences Academic Press (SSAP), is the mastermind behind the project. In addition to defining the key missions to be accomplished by it and setting down the basic parameters for the project's execution, as the work has unfolded, Mr. Xie has provided critical input pertaining to its every aspect and at every step of the way. Thanks to the deft coordination by Ms. Yanling Li, all the constantly moving parts of the project, especially those on the SSAP side, are securely held together, and as well synchronized as is feasible for a project of this scale. Ms. Jing Gao, unfailingly diligent and meticulous, makes sure every aspect of each Chinese manuscript meets the highest standards for both publishers, something of critical importance to all subsequent steps in the publishing process. That high-quality, if also at times stylistically as well as technically challenging, scholarly writing in Chinese has turned into decent, readable English that readers see on these pages largely thanks to Ms. Fan Liang, who oversees translator recruitment and translation quality control.

Ten other members of the SSAP staff have been intimately involved, primarily in the capacity of in-house editor, in the preparation of the Chinese manuscripts. It is time-consuming work that requires attention to details, and each of them has done this and is continuing to do this with superb skills. They are, in alphabetical order, Mr. Jihui Cai, Ms. Xiaojun Liu, Mr. Wenwu Ren, Ms. Xiaolin Shi, Ms. Yuehua Song, Mr. Genxing Tong, Ms. Dan Wu, Ms. Dongmei Yao, Ms. Wei Yun and Ms. Qiong Zhou. In addition, Shouguang Xie and Yanling Li have also taken part in this work.

Ms. Qiong Zhou is the SSAP in-house editor for the current volume.

Our appreciation is also owed to Ms. Yan Li, Mr. Ning Chai, Ms. Lei Wang and Ms. Yi Xu from Springer's Beijing Representative Office. Their strong support for the SSAP team in various aspects of the project helped to make the latter's work that much easier than it would have otherwise been.

We thank Mr. Yong Zhou for translating this book and Mr. Leeor Schweitzer for his work as the polisher.

Last but certainly not least, it must be mentioned that funding for this project comes from the Ministry of Finance of the People's Republic of China. Our profound gratitude, if we can be forgiven for a bit of apophasis, goes without saying.

Social Sciences Academic Press
Springer

Preface

The city is one of humanity's greatest inventions. Human beings have achieved enormous socioeconomic progress in modern times by focusing on cities. Many human civilizations began and developed with an urban core. Two hundred years ago, urban residents accounted for less than 3 % of the world's total population, but by 1900 that number had gone up to 13 % and spiked up to 50 % by the beginning of the twenty-first century. By 2050 70 % of the world's total population is expected to live in cities. Cities' rapid development, starting from nothing, follows similar trends in human development, ideals, and other changes. Cities serve as a ladder for human advancement. The process of urbanization is encapsulated primarily by the continuous human migration into cities. Joseph Eugene Stiglitz, an American economist and recipient of the Nobel Memorial Prize in Economic Sciences, predicted that high-tech development in the United States and urbanization in China will be the two primary factors shaping the world in the twenty-first century. Strong economic growth in China that began with the reform and liberalization policies has greatly promoted urbanization. Urbanization in China rose to 53.7 % in 2013, up from 17.9 % in 1978, demonstrating an average annual increase greater than 1 %. China has long been the world leader in urbanization, both in regard to net annual increase of urban population and the total urban population. China saw its permanent urban population increase to 730 million in 2013, up from 170 million in 1978. It is expected that about one billion people will live in Chinese cities by the end of 2030. By that time there will be 221 Chinese cities with a population of over one million (currently there are only 35 cities that are so large all across Europe). Of the 221 cities, 23 will have a population of over 5 million, and 8 cities will have a population of over 10 million, according to forecasts from McKinsey & Company. During this period, 30 million Chinese, a group equivalent to Canada's population, will join the middle class each year. Important changes will occur in the global production market when the size of China's middle class overtakes that of the United States and China becomes the world's largest consumer market. China's rapid urbanization leads to a massive consumption of resources, rapid expansion of the secondary and tertiary sectors, and the corresponding issues with employment, housing, transportation, and social security. It further forces lifestyle changes and carries huge environmental impacts.

China's socioeconomic development pattern has now entered a critical transformation period. Accordingly, urban development has also shifted and entered a new stage, a period of urbanization-oriented industrialization, opposite the approach used in the past. Transformation of the economic growth pattern and of the urbanization process stand as two approaches to a singular mission. China is now focused on economic restructuring, mainly the increase of domestic demand through urbanization, in order to accelerate these transformations. China's total population is greater than the population of all member countries of the Organisation for Economic Co-operation and Development (OECD) combined. An average annual increase of 1 % in China's urban population means that over 10 million people, greater than the total population of many European countries (only 15 of over 40 European countries had a population over 10 million in 2006), would migrate into cities from rural areas each year. Urbanization of the Chinese population is a strong, endogenous force upon which the socioeconomic development of China must rely. The transformation of both the economic growth pattern and of economic restructuring is dependent upon urbanization levels regardless of what those levels may be. Transformation and development planning has centered on urbanization so as to fuel comprehensive socioeconomic improvements in a China that has entered a new stage of urbanization.

China has made significant achievements in urban development since 1978. These include the rapid expansion of cities, strengthening urban economies, further optimization of industries, improvements to urban infrastructure and environment, continuous enhancements to urban services, improved urban construction and management techniques, and booming public services. These steps have aided China in working toward its goal of building harmonious cities. Chinese cities have progressed in differing ways since the beginning of China's reform and liberalization policies, although most have shifted from traditional city structures based on self-sufficient agricultural economies to modern city structures based on industrial market economies. Establishing a market economic system requires the introduction of capitalism, where resources are distributed on the basis of supply and demand calculations and attempts to increase one's economic advantage. This logic has radically changed an ancient China within several decades, one notable change being the emergence of many modern cities. Industrialization and the market economy have fueled China's urbanization but have also brought about enormous negative and sometimes unpredictable effects. As rural towns are dominated by factories, traditional houses built along small rivers and the many accompanying bridges, once a traditional landscape in southeastern China, have been replaced by high-rise buildings made of concrete. While enjoying certain elements of the urban lifestyles, people face increasingly severe urban problems heralded by the rapidly growing population such as tight housing supplies, difficulty getting a job, traffic congestion, environmental degradation, and frequent public safety incidents. Although memories of the Great Smog of London in December 1952 and of the photochemical smog of Los Angeles in the 1970s may have faded, the heavy smog that occurred in the city of Beijing and the surrounding areas in the winter of 2013 remains fresh in the public's mind. This smog covered 1.3 million square kilometers, lasted 3 weeks, reached a particulate matter (PM) 2.5 concentration in excess of 500 μg/m^3 in a number of cities, and affected over 800 million, an irregular magnitude on all counts that shocked the world and was referred to by an expert as "an incident worth recording in the world's history of pollution." This pollution incident has raised the public's attention to air quality to an unprecedented level. In a report on air pollution in cities worldwide, published by the World Health Organization in 2011, of the 1,083 cities in 91 countries, the 28 provincial capitals of China all ranked below 900. Smog in China results from multiple factors, top among which is the high energy consumption per unit of GDP (energy density). China's energy density is five times higher than Japan's, four times higher than Germany's, and three times higher than the United States'. China experienced a GDP growth of 9.2 % in 2011 and an energy consumption increase of 7 % during that same time, placing its total at 3.5 billion tons of standard coal equivalent, or 46 % of the world's total energy consumption. China's economy is expected to double by the end of 2020, but a similar doubling of its energy consumption is entirely unsustainable. The Chinese public has instead embraced the concept of socioeconomic development geared toward an ecological civilization. This evolving concept dispenses sharp criticism against the traditional GDP-centric model of modernization. Developed countries have begun a second phase of modernization scrutinizing traditional development that aimed at nothing but GDP growth. At the invitation of the United Nations Educational, Scientific, and Cultural Organization (UNESCO), the late French economist François Perroux published a book titled *A New Concept of Development* in 1983 to propose a people-centric, generalized, and local-based development concept that pays particular attention to the value of cultures. Indexes focused on green and people-centric data have now become basic measures for socioeconomic development that are generally accepted in the international community. As developed countries shifted toward information-based economies, traditional industries were moved to developing countries. As a state in the periphery of the global economy, China became an ideal destination for these industries since it was then only in the early stage of industrialization, only a short time after it rolled out its reform and liberalization policies. As a result, China became "the world's factory floor." While bringing China prosperity and advances, this shift has led to a series of negative effects, stemming from the overemphasis on development. Over time China has come to realize that urbanization centered around industry is unsustainable. China's urbanization trends have so far been deeply affected by GDP-centric policies, industrial emphasis, and capitalistic principles borne from a market economy. The foundation of the market economy in its purest form is a materialistic strategy that aims at nothing but GDP growth. "How can we adjust our society and culture to our industrialization?" This was how M. A. Sinaceur, a former UNESCO official, described the overemphasis of industrialization in the old development concept in his preface to the aforementioned book by François Perroux. The old development concept equates development with growth, equates growth with GDP increase, and hence uses GDP growth rate as the primary measure for development. Such a development concept is built upon material logic in an extreme form, which will ultimately lead to consequences such as converting residential areas into industrial facilities or their components, such as sites for obtaining, storing, and disposing of raw materials; severe damage to the environment and the livability of our surroundings; and the eventual impossibility of the reproduction of material goods. Chapter 1 of this book will explain how the current global ecological crisis has much to do with the industrialization of civilization since the nineteenth century. It is impossible to abate the environmental crisis under the framework of an industrial civilization. A transition to an ecological civilization is necessary to emerge from the environmental crisis. China will undergo the largest-scale urbanization in human history within the next decade, so that its decisions on urban development will affect the entire world's urban future. What will desirable Chinese cities be like? How should China finish the urbanization process? These are some of the fundamental questions important to shaping urbanization, and this book seeks to answer and explore these questions. The Chinese government has made a firm commitment to the international community to transition toward an ecological civilization. This transition involves China's future modernization; the transformation of entire social systems; changes to living, employment, and consumption habits; and massive societal reorganization. The Chinese government's long-term effort in search of optimized development patterns has culminated in a set of values that largely overlap with those of an ecological civilization. China has been engaging in environmental protection ever since it sent a delegation to the first ever United Nations Conference on the Human Environment held in Stockholm, Sweden, in 1972. The State Council, which acts as China's cabinet, held the first national conference on environmental protection in 1973, at which time it announced a 32-character guideline for environmental protection. After Agenda 21 was passed at the United Nations Conference on Environment and Development (UNCED) held in Rio de Janeiro, Brazil, the Chinese government established sustainable development as a national strategy for modernization and published *China's Agenda 21: White paper on China's Population, Environment and Development in the 21st Century*. The World Summit on Sustainable Development was held in 2002, just before the Chinese government announced that it would work under the Outlook on Scientific Development protocol in order to accelerate historic changes in environmental protection efforts and develop new approaches for building a resource-efficient and environment-friendly society. The 2012 United Nations Conference on Sustainable Development known as Rio +20 offered the Chinese government the opportunity to make clear its position that "We must prioritize ecological progress and incorporate an ecological perspective in all aspects and processes surrounding economic, political, cultural, and social progress. We must work hard to build a beautiful country, and achieve lasting sustainable development in the Chinese nation." A review of environmental protection efforts in China shows that the primary advances were achieved through value changes and the greatest changes were born of development concepts. Environmental protection across China has changed in important ways, including both theories and practices, since the 11th Five-Year Plan period. The ecological civilization concept already plays a leading role in guiding China's socioeconomic development. The Chinese government has announced that "We have to understand that to protect the environment is to preserve our productivity and to improve the environment is to develop our productivity. Such concepts should be deeply rooted." China has committed to consciously promote a green, sustainable, and low-carbon development pattern, and it will not sacrifice the environment for temporary economic growth. The Chinese government will prioritize resource conservation, environmental protection, natural restoration, and economic development achieved through environmental protection. China is the only country that has defined what is known domestically as "ecological red lines." The *National Ecological Red Line – the Technical Guide to Setting the Base Line for Ecological Functions (Trial)* was published in February 2013. This red line was adopted as national policy following the red line establishing 120 million hectares of farmland. The Chinese government has given top priority to setting ecological red lines in efforts toward the environmental protection administration reforms and in order to work toward systems oriented about an ecological civilization. At a study session with members of the Political Bureau of the Communist Party of China (CPC) Central Committee held in May 2013, Chinese President Jinping Xi stressed the implementation of the major function-oriented zone (MFOZ) strategy and pushed the responsible governmental departments to accelerate the process and to set ecological red lines in strict accordance with developmental priorities, emphases, restrictions, and prohibitions based on the zone-specific major functions that can then be properly adhered to. Xi also said that the ecological red line concept should be firmly established, with relevant punishments for those crossing the red line so as to ensure environmental protection. These statements underscore the Chinese government's resolution to develop toward an ecological civilization. The ecological civilization concept has already been incorporated throughout relevant systems, policies, laws, and local governments' urban development plans, as part of the socioeconomic development assessment system, and into corresponding accountability systems, and is currently being added in those places where it is not yet embedded.

Environment-friendly cities should be built such that natural and cultural ecosystems constitute a harmonious whole. Cultural elements unique to a given area form the soul of environment-friendly cities, therefore cultural inheritance should be highly respected. The Chinese government approved the National New Urbanization Plan (2014–2020) at a national conference on urbanization in December 2013. This plan cites two inseparable themes, ecology and humanity. A green civilization cannot be inhumane. The pitfalls of human nature serve as a backdrop to ecological disasters and urban problems. To build environment-friendly cities across China, how can we integrate ecological civilization concepts into all elements of urbanization? How can we find an approach to building environment-friendly cities that respects China's style? At the same urbanization conference where the National New Urbanization Plan was presented, a proposal was made that urban development should prioritize indigenous culture, develop fascinating sites with important cultural roots, and support local heritage. Both proposals address the primary issue regarding urbanization in China. Since urbanization is intertwined with culture, building cities that captivate using the local culture represents the highest standard for urban development and the preferred method for building environment-friendly cities. There is a Chinese saying that proclaims: it takes a decade to nurture a seedling into a tree, but it takes even longer to make a man of oneself. Urban development must be people oriented and must base itself upon continuous improvements in the populations' cultural literacy. By focusing on protection of cultural resources and improvements in people's cultural literacy, Chinese cities can develop toward an ecological civilization and become environment friendly.

Beijing, China
June 28, 2014

Jingyuan Li

Contents

Contributors

Guohua Chang Lanzhou City University, Lanzhou, China

Tianpeng Gao Lanzhou City University, Lanzhou, China

Lingfeng Kang Lanzhou City University, Lanzhou, China

Juheng Li Lanzhou City University, Lanzhou, China

Juke Liu Lanzhou City University, Lanzhou, China

Jiaxing Lu Lanzhou City University, Lanzhou, Gansu, China

Guoquan Qian Gansu Research Institute for City Development, Lanzhou, China

Weiping Sun Center for Social Development, Chinese Academy of Social Sciences, Beijing, China

Cuiyun Wang Lanzhou City University, Lanzhou, China

Dingjun Wang Lanzhou City University, Lanzhou, China

Taichun Wang Lanzhou City University, Lanzhou, China

Yongzhen Wang Lanzhou City University, Lanzhou, China

Tongjin Yang School of Public Policy and Management, Guangxi University, Nanning, China

Gang Zeng The School of Resources and Environmental Science, East China Normal University, Shanghai, China

Yongli Zhang Business College, Northwest Normal University, Lanzhou, Gansu, China

Youyi Zhao Lanzhou City University, Lanzhou, China

About the Authors

Jingyuan Li is a Full Member of the Chinese Academy of Social Sciences (CASS), Deputy Director of the CASS Academic Committee on Literature, History and Philosophy and Director of the Cultural Research Center, CASS. He is also researcher and doctoral student advisor in the Institute of Philosophy, CASS, where he formerly served as deputy director. Li has published widely in epistemology, Marxian philosophy and ethics. His representative publications include *A Study of our Understanding of Pre-history* (1989), *Genetic Epistemology* (1991), *Marxian Philosophy and Modern Times* (1991), *Eternal Charm* (1992), *Prolegomena to Methodology of Mao Zedong Theory* (1993) and *New Directions for Cultural Development in the New Century* (1999).

Tongjin Yang is the philosophy professor at the school of public policy and management, Guangxi University. Yang specializes in environmental ethics, environmental philosophy and applied ethics. His representative publications include *The Deepening of Environmental Protection* (1999), *Environmental Ethics – Global Discourse, Chinese Perspective* (2007), *The Ecological Turn of Modern Civilization* (editor, 2007) and *Chinese Applied Ethics – Special Volume on International Ethics* (2013).

Acknowledgments

The book was written by a number of authors with support from multiple organizations. We would like to express our heartfelt thanks to the Social Development Research Center of the Chinese Academy of Social Sciences (CASS), Lanzhou City University, and a researcher in urban areas and planning at East China Normal University for their assistance in this project. Ms. Qiong Zhou of the Social Sciences Academic Press (China) was ever interested in the progress of this project and provided specific guidance on how to write this book. Dr. Wenzhen Hu, Deputy Director of the CASS Social Development Research Center, also provided help for writing this book. Once again, our gratitude goes to all of them.

This book was coauthored by

Tongjin Yang, Weiping Sun, Juke Liu, Dingjun Wang, and Gang Zeng for Chap. 1
Weiping Sun, Juke Liu, and Dingjun Wang for Chap. 2
Guohua Chang for Chap. 3
Lingfeng Kang and Youji Zhao for Chap. 4
Yongli Zhang and Jiaxing Lu for Chap. 5
Guoquan Qian and Cuiyun Wang for Chap. 6
Taichun Wang and Tianpeng Gao for Chap. 7
Guoquan Qian and Yuxia Wang for Chap. 8

Part I
The Development of Eco-cities in China: Theories

Chapter 1
Ecological Crisis, Eco-civilization, and Eco-cities

Tongjin Yang, Weiping Sun, Juke Liu, Dingjun Wang, and Gang Zeng

1.1 Ecological Crisis in Industrial Civilization

1.1.1 Ecological Crisis: A Challenge for Human Civilizations

Humanity has been making efforts to protect the environment since the 1960s. The United Nations (UN) held the first ever conference on the human environment in 1972. Despite decades of activity, the environmental crisis confronted by humanity is growing in magnitude and is less controlled than ever. While local improvements have flourished, the overall state of the global environment has deteriorated. The additional predicament of climate change raises the environmental issue to the level of one of the greatest challenges for human beings in the twenty-first century. The

T. Yang (✉)
School of Public Policy and Management, Guangxi University,
Nanning, China
e-mail: 13718631990@163.com

W. Sun
Center for Social Development, Chinese Academy of Social Sciences,
Beijing, China
e-mail: swp1966@126.com

J. Liu • D. Wang
Lanzhou City University, Lanzhou, China
e-mail: ljk@lzcu.edu.cn; 519542428@qq.com

G. Zeng
The School of Resources and Environmental Science,
East China Normal University,
Shanghai, China
e-mail: gzeng@re.ecnu.edu.cn

J. Li, T. Yang (eds.), *China's Eco-city Construction*, Research Series on the Chinese Dream and China's Development Path,
DOI 10.1007/978-3-662-48153-0_1

ecological crisis facing humanity manifests itself around the world predominately in the following respects:

The Greenhouse Effect The global climate is warming year by year as a result of human activities. It is estimated that the average air temperature on Earth will be 5.8 °C higher in 2100 than in 1990. In other words, the average air temperature on Earth will be the highest it has been in 50 million years. A warming climate will raise the sea level, flooding many coastal and island countries and inviting more frequent storm surges, and also open the door for a series of natural disasters such as disease spread, depleted water resources, species extinction, pest infestation, and desertification of large stretches. Global warming has already become humanity's most dangerous and challenging headache.

Species Extinction There are about 25,000 plant species and 1,000 vertebrate species around the world that are on the verge of extinction. Human activities cause species to go extinct 100 to 1,000 times faster than they would naturally. Faster extinction simplifies ecosystems and creates imbalance. Once a species becomes extinct, it can never reappear.

Forests in Sharp Decline Forests are an important part of Earth's biosphere. Earth's total forested areas, which once covered 7.6 billion hectares, has reduced to 5.5 billion hectares in the nineteenth century and stands at less than 0.3 billion hectares today. Forests are disappearing at an average rate of 4,000 km^2 per year. Tropical rainforests are disappearing from around the world at an average rate of 20 ha per minute. If continued unabated, there will be no rainforest at all by the end of 2030.

Desertification Deserts already cover one fourth of the Earth's land surface. As many as six million square kilometers of land become deserts every year. Desertification affects the lives of about one sixth of Earth's population.

Water Scarcity and Pollution Over 100 countries suffer from water shortages. Three fourths of the rural population and one fifth of the urban population in developing countries have no access to safe and healthy drinking water. Water pollution intensifies the water crisis and is related to 80 % of all diseases and one third of human deaths.

Toxic Chemical Pollution About five million types of chemicals are produced around the world, including about 30,000 types harmful to human health and ecosystems, and over 500 types that may cause cancer, deformities, and mutations. Moreover, tens of thousands of new types of chemicals are invented every year, leading to a growing number of toxic chemical pollution incidents. These toxic chemicals trigger synergistic reactions, leading to ever greater harms to both the environment and human health.

Along with the rest of the world, China is afflicted with all of the aforementioned environmental problems. China is under even heavier environmental pressures and faces tougher environmental challenges than most other countries, since it is the most populous country and its economy has been growing rapidly for over 20 years.

Long-term pollution has rendered ecosystems across China exceptionally vulnerable. Desertification has already claimed 27.3 % of China's total land area and is increasing at an average rate of 2,460 km^2 per year. Human migration caused by desertification is particularly pronounced in some areas of Gansu Province and Inner Mongolia, where tens of thousands of people are forced to leave their desert-covered towns every year. China loses over 5 billion tons of soil every year, which contains the nutrient equivalent to 40 million tons of standard chemical fertilizers. China suffers one of the heaviest losses of biodiversity globally. 4,000–5,000 higher plants are endangered or nearly endangered, that is 15–20 % of all species in the country. The world's average level rests at 10–15 %. Known as the "Mother River" of the Chinese civilization, the Yellow River has long lost its vibrancy. The Yangtze River is also becoming China's second "yellow river."

There are over 2,000 natural lakes across China. Unfortunately, nearly 20 natural lakes have disappeared annually over the past five decades. Pollution has been found in 75 % of China's lakes. Qinghai Lake, the largest saltwater lake in China, is shrinking day by day. Since the 1960s, its water level has dropped by 3.7 m, and its area reduced by 312 km^2. There were as many as 78 rivers around Qinghai Lake, but most of them have dried up. Only a little more than 20 rivers, such as the Buha River, flow into Qinghai Lake, although total water flow is over 60 % less than it was in the 1950s. Dropping water level, desertification, aridity, and grassland degradation have destined Qinghai Lake to become a sandy or dead lake. The well-known Crescent Lake near the city of Dunhuang in Gansu Province was 1.47 ha in area and up to 9 m deep in the 1970s, but today it only measures 0.53 ha and less than three meters in depth. Crescent Lake will disappear in 30 years if no remedies are taken. China has been performing better than expected since it began reform and opening up in terms of annual economic and social indicators, in contrast with a lower than expected environmental performance. The ecological crisis threatens human civilizations and is imposing immediate dangers to the socioeconomic sustainability of China.

1.1.2 *The Root of the Global Ecological Crisis*

An unavoidable takeaway from observing the last 50 years of environmental protection work and the continued deterioration of the environment is that the environmental crisis is structurally embedded in industrialized civilization. Even the basic structures and mechanisms of industrial civilization are entangled with ecological crisis. It is impossible to eliminate the environmental crisis under the framework of industrial civilization.

The mainstream economic theory of industrial civilization proposes ambiguous resource ownership and business cost externalities as the two primary causes of environmental crisis. It goes on to suggest two natural solutions to environmental issues. First, clarify resource ownership such that every resource becomes the private asset of a particular economic entity. "The tragedy of the commons" can be avoided if every economic entity manages its private assets diligently. Second, internalize external costs. This demands of the government to require businesses to include costs associated with environment pollution caused by product production and use into the production cost. In order to internalize environmental costs, natural resources should be priced, because the existing pricing system does not reflect the cost of natural resources used for production, it only includes priced production inputs and transportation necessary for a product. Adding the cost of natural resource inputs and environmental externalities to a product will increase its price, thereby lowering consumer demand for the product. Dividing the ecosystem into parcels that can be distributed with clear ownership rights is not so simple, because the organic system functions as a whole. It is impossible to perfectly match the boundaries of environmental elements with demarcated property boundaries. Human beings are not completely rational; hence it is impossible to expect them to always behave rationally. The same reasoning follows for any economic entity. We should not expect that any person will care about the environment where future generations will have to live, even if he or she were capable of always managing his or her assets completely rationally, because doing so will not benefit their implied self-interest. Efforts to internalize external costs will also face many insuperable obstacles. Given the sporadic nature of environmental pollution, neither the government nor businesses are able to calculate the exact cost that must be internalized to counteract the pollution before production begins. Seemingly negligible pollution from many businesses will combine to produce accumulative effects, which will have environmental implications that are difficult to assess. Pricing the natural world involves comparison of generational preferences such that prices set by the current generation are unlikely to reflect the wishes of future generations. Pricing all of nature implies that natural resources are tradable and replaceable. This is a serious misinterpretation of nature.

The political tenants of industrial civilization are also unlikely to provide strong support for environmental protection. National sovereignty grants every country the absolute right to manage its internal issues. However, many environmental issues are global and require joint action. More nationalistic countries will resist cooperative environmental protection efforts. In addition, representative democracy will be under pressure from both the business community and the electorate. The business community can organize to apply pressure on the government to prevent environmental laws. The electorate cares more about their short-term benefits and generally pays little attention to government efforts geared toward long-term environmental issues that will only affect later generations. Elected representative may also act against the wishes of the electorate to please special interest groups.

Modern technology boosts industrial civilization. Technological advances will improve resource efficiency and slow down depletion. The potential benefits of technological advances for environmental protection have nevertheless been completely offset by the environmental pressures generated by explosive human population growth since the latter half of the twentieth century. The Jevons paradox tells us that a certain resource will be consumed and depleted at a faster rate if technology is introduced that improves that resource's use efficiency. This is due to sharp decreases in product prices when technological improvement are introduced, which increases demand for the product that requires the initial raw materials. Environmental issues are not purely technical, making it impossible for technological advances alone to solve them. Environmental protection is a complex social project that can effectively contain the ecological crisis in industrial civilization only through comprehensive institutional reforms.

The basic values of industrial civilization will never be entirely compatible with environmental protection. Industrial civilization views nature mechanically, treating it like clockwork. It follows that connections between the components of this machine are mechanical, and a general understanding of this machine can be achieved by understanding its parts (following reductionist epistemology). We are able to disassemble and reassemble this machine at will as long as we understand all of its parts and the mechanical connections between them. This mechanical model of nature provides humanity with a reliable basis for understanding and controlling nature. Many great modern scientific achievements relied on this world view. However, viewing nature as directly compatible to mechanical structures introduces many flawed conclusions. The perspective ignores the organic and dynamic characteristics present in nature as a single whole and the intricate, organic connections between its many components. Assuming that nature is a simple machine that can be arbitrarily disassembled and reassembled and that consists of parts with fixed, measurable, and controllable interconnections leads humanity to believe that it can conquer nature. Industrial civilization's attempt at controlling and conquering nature has become one of damaging and robbing nature because humanity underestimates nature's inherent complexities and overestimates human ability and control. The damaged nature is the root of civilization, and attempting to conquer it destroys the environment and the home for human beings.

The mechanical view of nature deprives it of meaning and value by reducing it to a set of objects that can be expressed with numbers. This delegates all meaning and value to human societies and the human soul. Humanity, comprising of souls and rationality, and nature become completely different beings (an understanding expressed by axiological dualism). Axiological dualism distinguishes between humans and nature with regard to value and further separated the humanities (research of human phenomena which are valuable and meaningful) from natural sciences (research of natural phenomena, which are meaningless and valueless according to axiological dualism).

The mechanical view of nature and axiological dualism has laid a firm philosophical basis for anthropocentrism, which is widely accepted in industrial civilization. Anthropocentrists see differences between humanity and nature and promote humanity as playing the central role, but ignore the inherent connections between nature and humanity and deny an intrinsic value of nature. Anthropocentrists also believe that the central role of humanity is manifested by conquering and controlling nature. Such egocentric values are highly compatible with individual egoism, group egoism, and nationalism. They are all to blame for the environmental and social crises in the era of industrial civilization.[1]

---

[1] Yang [1].

1.2 Eco-civilization: A Necessary Path to Eliminating the Ecological Crisis

Based on historical example and the current state of affairs, partial modifications and improvements to the economic development pattern, political systems, technology, and values under a basic framework of an industrial civilization can only temporarily alleviate the pressure of the ecological crisis. Only by developing toward an eco-civilization can humanity eliminate this crisis as a threat to humanity. Shifting the paradigm of a civilization is necessary for humanity to eliminate the ecological crisis.

1.2.1 Basic Concept of an Eco-civilization

The eco-civilization is a developing paradigm. It represents an advanced developmental stage of human civilization that follows the industrial civilization. The eco-civilization is primarily characterized by a focus on harmony between humanity and nature. The economy is developed in eco-civilizations using an eco-economy, which considers the human economic system as part of the natural ecosystem rather than forcing the natural ecosystem into the human economy. The eco-civilization prioritizes the interests of the entire human race and advocates global governance and world citizenry. Science and technology are not viewed as tools for humanity to conquer nature but as assistants in restoring ecosystems and realizing harmony between humanity and nature. An organic view of nature is valued in an eco-civilization, highlighting the entirety of nature and its intrinsic value. The values of an eco-civilization do not only cover human rights but further emphasize care and responsibility and advocate harmony and rational consumption.

The eco-civilization views nature organically, accepting it as an integral whole and recognizing the connections between all its parts as organic and inherent. This view of nature maintains that nature has an intrinsic value. Nature is understood to hold abundant and evolving value and to provide the source of life and meaning. The value of humanity becomes no more than an extension of the value of nature. The unique value and central role granted to human beings relies on the fact that they are the only beings in nature with a moral consciousness and are able to understand nature's potential to create, maintain, and improve living beings and are able themselves to maintain and improve nature's potential by controlling their behavior toward nature with moral ideals. Maintaining an ecological balance on Earth is regarded as a key path to realizing the value and central role of humanity within and eco-civilization, and the harmonious development of humanity and nature is recognized as an intrinsic spiritual need and a new way of existing for humanity.

The eco-civilization must be sustainable. Humans' only home, the Earth, provides us with limited resources and a limited ability to absorb pollutants generated by us. Humanity's economic system is only part of the greater ecosystem and can become sustainable when supported by the natural ecosystem. The economic system of the eco-civilization should run such that it is within the carrying capacity of the ecosystem, achieving positive interactions between the two systems and a harmonious development of both. Reducing and preventing pollution and improving the resource utilization should be reflected in every aspect of economic activities. The social, economic, and ecological benefits of economic activities can be realized by developing clean technologies, introducing industrial ecology into production, and advocating for eco-agriculture.

Highlighting sustainable development is an outstanding characteristic of an eco-civilization. Sustainable development is impossible without sustainable natural and social environments. In order to leave future generations a sustainable natural environment, the eco-civilization will undertake a sustainable resources development strategy, which will include: preserving biodiversity and increasing the natural capital reserve and its percentage of national wealth; improving resource use and enhancing the Earth's carrying capacity through technological innovation; limiting the use of organic resources on Earth such that they are able to regenerate naturally; considering the rate at which nonrenewable resources are exhausted and endeavoring to find alternatives before they are exhausted; and developing clean energies such as hydro, solar, and wind power to reduce dependence on polluting fossil fuel energy sources such as crude oil and coal. The eco-civilization will strive to leave future generations a sustainable social environment by adopting measures to make a more equitable society, which include: building a set of systems that better satisfy the basic needs of people and letting people exercise their basic rights, keeping the size of the human population within a proper limit, improving the physical and spiritual characteristics of humanity, providing better education, and advocating for more ecological consumption choices and lifestyles. The eco-civilization will be required to adhere to a new level of justice. Justice in an eco-civilization includes equity between humanity and nature, between all people of the current generation, and between the current generation and future generations. Equity for humanity and nature requires balancing the needs of ecosystems and social systems by harmoniously developing humanity and nature such that the ecosystem remains balanced and stable, while human needs for survival and development are satisfied. The basic human needs in an eco-civilization will be satisfied in the same way as basic human rights are protected. This aims to satisfy the basic needs of all people such that they are able to further develop, which is the quintessential meaning of an equitable society.

Eliminating poverty worldwide is an important aspect of building an eco-civilization. The extremely rich, developed countries have created environmental pressures around the world and further degraded the environment by spreading poverty, especially in nationalistic countries. Poor countries typically seek rapid economic growth at the cost of the environment. These countries are forced to export their natural resources at low prices to obtain foreign currencies needed to repay debts. Poor countries lack the technologies and the budgets for programs needed for environmental protection. People of the lowest income group, especially those who live in cities, are vulnerable to harms from toxic waste and hazardous chemicals. They often work in heavily polluted environments and are subject to the greatest damages from environmental disasters. The huge gap between rich and poor further defies human ethics. The elimination of poverty worldwide must be on the agenda of any global environmental governance system within an ecological civilization. The vicious cycle of poverty leading to environmental degradation leading to poverty must be broken. Justice will be reconceived as global environmental justice in the eco-civilization. An unjust international order reproduces and worsens environmental injustice. The most advantaged groups are able to reap the greatest benefits from the environment, whereas the least advantaged groups are subject to the most environmental disasters. All people are entitled to basic environmental goods according to the environmental justice lens. The right to the environment is treated as a basic human right. Every country is responsible to protect its citizens' right to the environment. Realizing global environmental justice is a responsibility of governments and citizens alike. Global environmental justice is rooted in public consciousness. Citizens worldwide will proactively demand their governments work with other countries to create an international order that buffets global environmental justice only when most of them have realized the importance of the global environment and have assumed their own responsibilities toward the cause. The eco-civilization will also introduce holistic thinking. All living things on Earth are members of one big family. They interact with each other and also constitute an integral, organic whole along with the Earth. As a late comer to this big family, humanity cannot exist without support from ecosystems and from other living things even if humans have used their intelligence to secure a large space for their survival. As human activities increasingly penetrate into every corner of the Earth, the fate of humanity becomes ever more intertwined with that of the other members of this big family. Extinction of other species spells an eventual death for the human race. As the strongest living beings and the only species with moral consciousness on Earth, humanity should assume the responsibility for protecting biodiversity and the Earth. Holistic thinking emphasizes the organic connections between humanity and nature and the necessity of human beings acting as a whole to address the environmental crisis. The well-being of individual nations is becoming increasingly connected as globalization accelerates. Environmental pollution does not recognize national borders. No nation is able to solve global environmental issues facing human beings on its own. Environmental efforts by any one nation, or even by a small grouping of nations, will come to naught without simultaneous corresponding actions by all other nations. Global action of greater depth and breadth should be taken in an eco-civilization. A more equitable political and economic order should be created, and a more effective system for drafting and enforcing environmental law should be built through institutional innovation on the international level. The most important themes of human societies – peace, development, and environmental protection – should be organically and holistically combined with sustainability into any development strategy. This practice will enable simultaneous achievement of peace, economic development, and environmental protection.

The eco-civilization is built upon previous civilizations. The eco-civilization is broadly oriented toward the harmonious development of humanity and nature and requires coordination and sustainable development of the human population, economies, resources, and the environment. The eco-civilization exists parallel to all the other human civilizations and emphasizes how civilized human beings interact with nature. The eco-civilization is essentially consistent in spirit with all the other human civilizations. First, equity between humanity and nature is valued. Humanity is a part of all other living beings in nature, that is, humanity is a child of nature. Human beings must never regard themselves as the masters of nature in an attempt to conquer it. Otherwise, human beings will be brutally punished by nature for their greed. Second, the unity of humanity and nature as opposites is exposed. There is a dialectical relationship between humanity and nature – they are unified yet at the same time opposite. Science and technology may both benefit and harm human beings. Human beings must obey the laws of nature, respect nature, and protect the Earth by controlling their desires. Only these steps can ensure socioeconomic sustainability. Third, intergenerational equity is valued. Human beings must account for the natural ecosystems sustainability when exploiting natural resources. The current generation is obligated to leave future generations a good environment for their survival and development and to achieve harmony between humanity and nature, as well as between living and nonliving things. Human beings can get out of the current ecological crisis only by building an eco-civilization and can create a more successful civilization by smoothly finishing a transition from an industrial civilization to an eco-civilization.[2]

1.2.2 *Metrics for an Eco-civilization*

Some international organizations, academic bodies, and scholars have made great progress in recent years toward quantitatively analyzing progress of an eco-civilization. Universally accepted metrics for this purpose have yet to be developed.

1.2.2.1 Typical Metrics

The Driving Force–State–Response (DSR) Framework Developed by the United Nations Commission on Sustainable Development (UNCSD)

The most influential metric today for an eco-civilization is the DSR framework developed by the UNCSD. Created in 1996, this framework was designed using the DSR model and initially included 134 social, economic, environmental, and institutional indicators. Strong logical ties are established between these indicators, which highlight the cause-and-effect relationship between environmental stresses and degradation and damage justifying closer study. The driving force indicators are

[2] Yang [2].

used to monitor human activities, processes, and patterns that affect sustainable development; the state indicators monitor the state of each system during sustainable development; and the response indicators correspond to policy choices. This model suffers from some deficiencies. The model struggles to demonstrate its internal logic in dealing with social and economic indicators. The indicators are categorized in an ambiguous manner, and the framework is broken down with uneven granularity. It is structurally unbalanced and lacks a clear thread for revealing the behavioral essence of sustainable development.

OECD Indicator Sets

The OECD indicator sets fall into three categories: the OECD Core Set of environmental indicators, OECD sets of sectoral indicators, and environmental accounting indicators. The Core Set, about 50 indicators, covers issues that reflect the primary environmental concerns in OECD countries. These indicators are classified into indicators of environmental pressure, environmental conditions, and society's responses using the pressure–state–response (PSR) model developed by Canadian statistician Anthony Friends, which are mainly used to track and monitor trends in environmental changes. This framework's primary advantage is the linear relationship between human activities and the environment that it reveals. The OECD sets of sectoral indicators focus on specific sectors and are classified into indicators of sectoral environmental significance trends, interactions with the environment (including positive and negative effects), and related economic and policy considerations, based on an adjusted PSR model. Environmental accounting indicators focus on physical resource accounts related to sustainable management of natural resources and environmental expenditure. Examples of these indicators include the intensity of natural resource use and the level and structure of pollution abatement and control expenditure. The OECD selected a small set of key environmental indicators as a tool for OECD work and for public information and communication by the OECD. The key environmental indicators aim to increase the public's environmental awareness and draw public and decision makers' attention to key environmental issues.

United Nations Statistics Division (UNSD) Indicators of Sustainable Development

Developed by the UNSD in 1994, the framework for indicators of sustainable development (FISD) is based on modifications to the framework for environmental data production (FDES). The FISD has 31 indicators and is based on the PSR model similarly to the framework developed by the UNCSD. The framework's primary advantage is that indicators in four categories were designed based upon issues in nine dimensions from *Agenda 21*, including economic issues, social and statistical issues, air and climate, land and soil, water resources, other resources, waste, human settlements, and natural disasters. The indicators are in the following four categories: social and economic activities and natural events, which correspond with pressure indicators; environmental impacts of activities and events and inventories, stocks, and background conditions, which both correspond with state indicators; and responses to environmental impacts, which correspond with response indicators. The FISD has some deficiencies including: 88 indicators, which just as for the UNCSD metrics, is too many; more environmental indicators than social and economic indicators; a lack of institutional indicators; and inconvenient use.

The UNSD System of Environmental–Economic Accounting (SEEA)

The UNSD noticed that the System of National Accounts (SNA), which was introduced in 1986, had two major deficiencies – a lack of consideration for the scarcity of natural resources and for worsening environmental quality. The UNSD began developing a new national economic accounting system – the System of Environmental–Economic Accounting (SEEA) – in 1993. The SEEA aims to evolve GNP/GDP into green GNP/GDP using a sustainability perspective. The SEEA is a massive data system designed to research how economies relate to the environment. The project faces implementation limitations such as availability of data. This system is also imperfect. Despite the setbacks, the SEEA still has decisive effects on the design of the ecological and environmental accounting systems in countries around the world.

The SCOPE Indicators of Sustainable Development

These metrics were developed by the Scientific Committee on Problems of Environment (SCOPE) in 1995. The SCOPE employed a method that was almost identical to the indicator consolidation method used by Albert Adriaanse of the former Dutch Ministry of Housing, Spatial Planning, and Environment in a 1993 book titled *Environmental Policy Performance Indicators: A Study on the Development of Indicators for Environmental Policy in the Netherlands*. This comprehensive, consolidated framework covers the environment, natural resources, natural systems, and water and air pollution to include 25 indicators of sustainable development. This framework's primary advantage is its consolidated set of indicators compared with other frameworks' unwieldy number of indicators. It nevertheless faces the same problems as the UNSCD and OECD metrics.

The World Bank's Indicator of Sustainable Development: The Genuine Savings Rate (GSR)

The World Bank presented the concept of genuine savings – net savings minus the value of resource depletion and environmental degradation in a country – in a report titled *Monitoring Environmental Progress* in 1995. Genuine savings has since been used as a new measure of the state and potential of a national economy. Research on genuine savings starts at GNP same as a traditional macroeconomic indicator. Gross savings for a country in the reference period equal GNP minus public and private consumption. Net savings equal gross savings minus the depreciation of product capital. Net savings focuses on product capital and does not represent development sustainability. Genuine savings take an extra step and are equal to net savings minus depreciation of natural capital and damages to the national economy caused by pollution. Genuine savings represent the total production amount that a country is realistically able to lend and invest in productive capital. The GSR can guide policy by suggesting, if it is negative, that the studied economy is ultimately headed toward reduced public welfare and suffers from unsustainable policies. The GSR is a framework for comprehensive and dynamic assessment of human development that emphasizes current connections between primary human development aspects and characteristics. The GSR is able to more accurately measure the real wealth and developmental capacity of a country than GDP is theoretically able to. Data is processed using monetization and addition methods to calculate genuine savings. Genuine savings enable monitoring and comparison of sustainable development processes in different countries. This method uses some concepts that require further refinement and research on how to quantify them, such as social capital. This indicator is more difficult to use because of the high technical requirements for the detailed calculation of the GSR. The indicator also overlooks spatial differences and basic conditions in different regions with respect to sustainable development.

Eurostat Indicators of Sustainable Development

The Eurostat published a report titled *Measuring Progress towards a More Sustainable Europe* that focuses on four pillars of the sustainable development strategy – economy, society, environment, and mechanism. A set of 63 indicators were designed, including 22 social indicators, 21 economic indicators, 16 environmental indicators, and six mechanism indicators. These metrics have the advantage that they are consistent with the policy classes on the CSD list and provide preliminary explanations for the main issues emphasized by the EU strategy. There is insufficient data available for most environmental variables, and the social indicators are regularly of a lower quality than the economic indicators.

The International Union for Conservation of Nature (IUCN) Barometer of Sustainability

The IUCN Barometer of Sustainability attempts to assess the state of humans and the environment and the progress toward sustainable development by measuring human and ecosystem well-being without submerging one in the other. The subsystem measuring human well-being includes 36 indicators in five dimensions including health and population, wealth, knowledge and culture, community, and equity. The subsystem measuring ecosystem well-being includes 51 indicators in five dimensions including land, water, air, species and populations, and resource use. These 87 indicators are averaged and weighted and then combined into the Human Well-Being Index (HWI), the Ecosystem Well-Being Index (EWI), and the Well-Being/Stress Index (WSI), respectively. The Barometer of Sustainability is a structural analysis program that assesses sustainable development and provides a comprehensive method for measuring the state of sustainable development, which applies at the local, regional, national, and international level. The weighting of the indicators requires researchers' subjective judgments and lacks scientifically shared criteria. In addition, the process of calculation is complex and incorporates unavoidable uncertainties.

The Index of Sustainable Economic Welfare (ISEW)

In order to measure public welfare in a country in a more comprehensive manner, Herman Daly and John B. Cobb developed the ISEW in 1989. Factors such as income distribution, environmental degradation, domestic labor, and depreciation of natural capital are adjusted and included in the accounting. This index attempts to address economic welfare issues overlooked by the SNA. It begins with consumption and further incorporates some environmental and distribution issues in the calculation of relevant indicators. Although the tool remains far from ideal when measuring economic welfare, the ISEW provides a better measurement than GDP. The index only takes a small number of resource and environmental factors into account.

The Genuine Progress Index (GPI)

The GPI is and index developed by Cobb et al based on the ISEW. The GPI's purpose is identical to that of the ISEW, creating a proper method for measuring economic welfare. It varies in that it focuses on assessing the effects, costs, and benefits of economic activities in over 20 aspects overlooked in the GDP indicator of economic development. The GPI measures welfare and genuine progress in a country more accurately than the GDP. Its accounts are developed using traditional accounting standards and represent a synthesis of many preexisting measurement systems. The GPI is now recognized as a new method for measuring economic development and may provide an alternative to the GDP. Data access, processing, and integration pose the primary challenges for progress with the GPI.

Metrics for Valuing Ecosystem Services

The metrics for valuing ecosystem services developed by Constanza et al in 1997 incorporate 17 ecosystem services over 16 biomes. These indicators aim to quantitatively assess the value of ecosystem services around the world. The research by Constanza et al reveals the market and non-market value of global ecosystems and fully analyzes their service value for humanity in a comprehensive manner for the first time. Global economists have since unleashed a wave of research on the value of ecosystem services. These metrics have the primary advantage of monetizing the functions of ecosystems that serve humanity and fully considering and calculating other ecosystem contributions to human socioeconomic systems. The conversion coefficients for using different biospheric ecosystem service functions are inaccurate when applied to local areas.

Emergy Analysis

Emergy analysis is an energy-centric systematic analysis method developed by H. T. Odum in 1996 that makes use of ecosystem and economic characteristics as well as the laws of thermodynamics. Emergy analysis converts different and incomparable energies in ecosystems into emergy (usually solar energy) to provide a common measure, thereby measuring and analyzing operating characteristics of ecosystems or eco-economic systems as well as the sustainability of their development. Emergy analysis makes up traditional monetary standards' deficiency in measuring the contribution of nature to economic development by simultaneously measuring human and environmental contributions to economic development. The use of a single conversion probability for multiple resources or products is, however, liable to be inaccurate. Emergy analysis also overlooks other limiting factors for economic sustainability.

Metrics for Sustainable Development Developed by the Sustainable Development Research Team at the Chinese Academy of Sciences (CAS)

The CAS developed a set of 208 indicators in 1999, which covers the overall, system, state, variable, and element layers in accordance with China's sustainable development strategy in the theoretical, structural, and statistical dimensions. The overall layer is intended to express the overall capacity of sustainable development and represents the overall progress and effects of implementing a strategy. The system layer consists of the production, development, environment, social, and intelligence support systems. The state layer is a relational structure that represents system behavior within each system statistically or dynamically. The variable layer uses 48 indicators to reflect the causes and motivations behind the behavior, relations, and changes of states. The element layer uses available indicators or groups of indicators that are measurable and comparable to directly measure the quantity, intensity, and rate of the variable layer.

Ministry of Science and Technology's (MOST) Indicator Set for China's Sustainable Development

The MOST built an indicator set for China's sustainable development in accordance with China's Agenda 21 released by the State Council (the Administration Center for China's Agenda 21 (ACCA21), 2004). This system comprises of the objective layer, reference layers 1 and 2, and the indicator layer. The descriptive and evaluative indicator systems are set on the indicator layer, comprising of 196 descriptive indicators and 100 evaluative indicators. This indicator set highlights the interactive, mutual, and mutually conditional cause-and-effect relationships between the overall concept of sustainable development and the indicators independently. These indicators provide a wide coverage that enables scientific assessment of the overall state of China's sustainable development. Problems arise in specific applications including overcomplexity of indicators, difficulty measuring and comparing regions using the same indicator, limitations of usable data, the likelihood of incomplete situational analysis, and one-sided conclusions. These research results lay a solid foundation for in-depth research on the eco-civilization's indicator set (Table 1.1).

Indicators used to assess progress toward an eco-civilization are increasingly mature and reliable, although they still suffer deficiencies, primarily in that they are only narrowly applicable (their application is limited by an area's or an institution's size), they are insufficiently dynamic (they lack effective z-temporal combinations making it difficult to achieve dynamic assessment), they fail to provide warnings and adjustments (research on indicator changes and trends goes unheeded), and they lack subjective indicators (some subjective indicators are overlooked because they are difficult to quantify). It is even more difficult to objectively and accurately assess areas with complex structures, low biodiversity, high human population density, and high material flow intensity using these indicators.

1.2.3 Global Development of Eco-civilizations

International organizations such as the UN play an irreplaceable role advocating, organizing, coordinating, and promoting progress toward eco-civilizations around the world. These organizations have opened up a new chapter in the global development of eco-civilizations by helping human societies agree on sustainable development.

Table 1.1 Overview of primary sustainable development and eco-civilization indicator sets

Type	Name	Description
Systemic	The DSR framework	Developed by the UNCSD, this framework consists of social, economic, environmental, and institutional indicators
	Sustainable Capacity (SC)	Developed by the CAS sustainable development research team, the SC includes 251 indicators covering the overall, system, state, variable, and element layers
	China's sustainable development indicator set	The MOST's ACCA21 built an indicator set for China's sustainable development according to China's Agenda 21 released by the State Council in 2004. This system comprises of the objective layer, reference layers 1 and 2, and the indicator layer. The descriptive and evaluative indicator systems are set on the indicator layer, including 196 descriptive indicators and 100 evaluative indicators
Societal	Human Development Index (HDI)	The United Nations Development Program (UNDP) combined education, life expectancy, and income status into a single measure of a country's human development for the first time in 1990, leading to the HDI
Economic	National Wealth (NW)	The World Bank proposed that the actual wealth of a country or region and dynamic changes over time to sustainable development capacity be assessed in the natural, productive, human, and social capital dimensions in 1995
	Sustainable Economic Welfare (SEW)	The SEW indicator was co-developed by the acclaimed economist Herman Daly and theologian John B. Cobb in 1989. This indicator incorporates income distribution and environmental degradation beyond average consumption levels, as well as factors ranging from the consequences of global warming to damage to the ozone layer.
	Green GNP	This is an accounting system co-developed by David Pearce et al in 1993 by modifying traditional national economic accounting methods. Green GNP equals the traditional GNP minus the consumption and depreciation of natural capital
Environmental	Environmental sustainability indicator set	This comprehensive set of 25 sustainable development indicators over environmental, natural resources, natural systems, and air and water pollution dimensions was developed by the SCOPE in 1995

(continued)

Table 1.1 (continued)

Type	Name	Description
Ecological	Ecosystem Services (ES)	This indicator was proposed by Lubcheno Constanza et al in an article published in *Nature* in 1997. The researchers took the lead in systematically estimating the value of services nature provides to humanity
	Ecological Footprint (EF)	The EF is a measure of humanity's use of natural resources and of life support services that nature provides humanity co-developed by Canadian ecological economists William Rees and Mathis Wackernagel

Source: Zeng Gang [3]

1.2.3.1 The UN Strategic Initiative for Sustainable Development

The United Nations Conference Declaration on the Human Environment (Stockholm Declaration) made in 1972 offers a systematic review of the ecological crisis in the era of the industrial civilization. The UN established the World Commission on Environment and Development (WCED) and chose Gro Harlem Brundtland, the former Prime Minister of Norway, to chair the commission in December 1983. The commission released a report titled *Our Common Future* in 1987, which coined the term "sustainable development." This document represented a major development of human thought on the environment and development by emphasizing the link between environmental protection and human development. Agenda 21 further emphasized and deepened theories of sustainable development when passed at the UNCED in 1992. A global sustainable development strategy was also proposed at the UNCED in 1992, opening the door to the eco-civilization. The UN held the World Summit on Sustainable Development (WSSD) in Johannesburg, South Africa, in 2002, asking countries to improve implementation of the quantified indicators specified in Agenda 21. Agenda 21 has stimulated research on the theories and practices of the eco-civilization by providing a global legal framework for its creation.

1.2.3.2 Responding to Global Climate Change with Low-Carbon Development

An Intergovernmental Negotiating Committee produced the text of the United Nations Framework Convention on Climate Change (UNFCCC) on May 22, 1992, which was then passed at the UNCED on June 4, 1992. The UNFCCC became the world's first ever international treaty that sought to comprehensively control emissions of greenhouse gases such as carbon dioxide, in order to address the adverse effects of global warming on human economies and societies. The treaty also provides a basic framework for international cooperation tackling global climate change. The UNFCCC went into effect on March 21, 1994. All UNFCC parties met once a year. The UNFCCC provides and generalized authoritative and comprehensive legal framework that laid a legal basis for international cooperation on climate change. The UNFCCC boasts 192 parties today.

The UNFCCC seeks to control carbon dioxide, methane, and other greenhouse gas emissions so as to "stabilize greenhouse gas concentrations in the atmosphere at a level that would prevent dangerous anthropogenic interference with the climate system." The treaty differentiates duties and procedures assigned to developed and developing countries. It requires developed countries to take specific measures to limit greenhouse gas emissions, as they are the major emitters, and funds developing countries for expenses needed to fulfill their duties under this treaty. Developing countries are only asked to provide lists of national greenhouse gas sources and sinks and of plans and measures, implemented or under development, to limit sources and sinks. The countries are not asked to comply with any legally binding limits on greenhouse gas emissions. The treaty has established mechanisms for providing developing countries with funds and technologies to aid in fulfilling their assigned duties.

The Kyoto Protocol was adopted as a supplement to the UNFCCC at the third session of the Conference of Parties to the UNFCCC (COP 3) in December 1997 in Kyoto, Japan. This protocol specifies greenhouse gas emission targets, country-specific duties and responsibilities, and implementation mechanisms. A resolution was formally passed on December 15, 2007, at the Bali Climate Change Conference urging negotiations on new arrangements for responding to climate change before the end of 2009. The participating nations adopted the highly anticipated Bali Road Map, which established a clear agenda for negotiations to be held by the end of 2009, including key climate change issues that were to be addressed. The Bali Road Map required developed countries reduce greenhouse gas emissions by 25–40 % by 2020. This road map is a milestone in promoting global development toward a low-carbon economy. The United Nations Environment Program (UNEP) decided on "Kick the Habit! Towards a Low-Carbon Economy" as the theme for the June 5 World Environment Day in 2008. Participants in the July 2008 G8 summit announced that they would seek to achieve at least a 50 % reduction of global greenhouse gas emissions by 2050 along with the other parties to the UNFCCC as a long-term goal. Developing a low-carbon economy was crystallized into consensus in the international community at the Copenhagen Climate Change Conference in 2009.

1.2.4 Eco-civilization in China

China, the world's largest developing country, enjoys the advantage of developing and industrializing after many other countries, allowing it to learn and beware of the pollution before restoration trend rampant in the history of industrialization in developed countries. China also faces severe population, resource, and environment conflicts, reflected by increasing occurrence of resources exhaustion, environmental degradation, and ecosystem damage, which is due to traditional economic growth patterns and ideological, technological, and institutional factors. China began changing its economic growth pattern in the 9th Five-Year Plan period and has since made great efforts, establishing a sustainable development strategy, developing toward an eco-civilization, and promoting concepts such as cleaner production, a circular economy, socialist new villages, eco-cities, an innovative country, a resource-efficient society, and an environmentally friendly society. Pursuing these strategic objectives and effectively implementing corresponding programs and plans fully demonstrate China's relentless efforts in developing toward an eco-civilization.

1.2.4.1 Sustainable Development Established as a National Strategy

China was the first country to release a national Agenda 21, *China's Agenda 21: White Paper on China's Population, Environment and Development in the twenty-First Century*, which was passed at the 16th executive meeting of the State Council on March 25, 1994.[3] The sustainable development strategy was included in the *Proposal for the Ninth Five-Year Plan on National Economy and Social Development and Long-Range Objectives for the Year 2010* adopted at the 5th Plenary Session of the 14th CPC Central Committee in 1995 and at the 4th Session of the 8th National People's Congress (NPC) in 1996. Sustainable development has since become a predominant national strategy for modernizing China in the twenty-first century. China submitted *The People's Republic of China National Report on Sustainable Development* to the United Nations Conference on Sustainable Development in June 1997. The former State Development Planning Commission (SDPC) developed the *Program of Action for Sustainable Development in China* in 2000, which outlined major fields and action plans for sustainable development in China for the twenty-first century.

[3] Qin et al. [4].

1.2.4.2 Building an Eco-civilization Under the Scientific Outlook on Development

The 1st Plenary Session of the 16th CPC Central Committee in 2002 made clear that "sustainable development capability will be steadily enhanced. The ecological environment will be improved. Resource use efficiency will be increased significantly. We will enhance harmony between man and nature to push the whole of society onto a path toward civilized development featuring production growth, an affluent life and a resilient ecosystem." These were among the major objectives for building a moderately prosperous society. The 3rd Plenary Session of the 16th CPC Central Committee required that a comprehensive, integrated, sustainable development concept be established to promote the overall development of society, economy, and humanity. This requirement has clarified the underlying role of the eco-civilization set against the history of other human civilizations.

The Scientific Outlook on Development was further explained in a report presented at the opening ceremony of the 17th CPC National Congress held in October 2007, which sought to "promote a culture of conservation by forming an energy efficient, resource efficient, and environmentally friendly industry structure, growth pattern, and consumption trend."[4] This outlook has ushered China into a new stage of eco-civilization construction.

The CPC Central Committee's Proposal on Formulating the 12th Five-Year Program (2011–2015) on National Economic and Social Development passed at the 5th Plenary Session of the 17th CPC Central Committee held on October 18, 2010. This document required that "the creation of a resource efficient and environmentally friendly society be accelerated, and the culture of ecological conservation be promoted." It emphasized that China's government authorities must increase awareness of the environmental crisis and establish green or low-carbon development, that they should improve incentives and restrictions with a focus on energy conservation and emissions reduction, and that resource efficient and environmentally friendly patterns of production and consumption should be established more rapidly to enhance the sustainable development capacity in light of increasing limitation of resources and the environment.[5]

1.2.4.3 Developing a Circular Economy

Developing a circular economy is a critical step for China to build an environmentally friendly society. In October 1993, the former State Environmental Protection Administration (SEPA) and the former State Economic and Trade Commission (SETC) jointly held the 2nd National Meeting on Industrial Pollution Prevention

---

[4]Hu [5].

[5]The 17th CPC Central Committee, *The CPC Central Committee's Proposal on Formulating the 12th Five-Year Program (2011–2015) on National Economic and Social Development*, http://news.xinhuanet.com/politics/2010-10/27/c_12708501.htm

and Control in the city of Shanghai taking "focus on prevention combined with control" as a theme and promoting cleaner production. The *Cleaner Production Promotion Law of the People's Republic of China (Draft)* was released in 1999. The *Cleaner Production Promotion Law of the People's Republic of China* became effective on January 1, 2003.[6] The circular economy concept has gained increasing acceptance since 2002. The former SEPA launched a number of pilot projects intended to develop a circular economy in certain areas. It picked the city of Guiyang and the province of Liaoning to be participants in a pilot program designed to develop a circular economy in May 2002. The Chinese government announced in 2003 that developing a circular economy was necessary to achieving sustainable development under the Scientific Outlook on Development. In 2004, it added that great efforts would be made to develop a circular economy all across society, including in enterprises, industrial parks, and cities.[7]

Developing a circular economy has recently become one of China's national strategies. The State Council has issued documents including *Some Opinions on Accelerating the Development of a Circular Economy* and the *Notice on Issuing the Comprehensive Work Plan for Energy Conservation and Emissions Reduction.* The National Development and Reform Commission (NDRC) has worked with the Ministry of Environmental Protection (MEP) to launch a national pilot program for developing a circular economy, conduct cleaner production auditing, and implement medium-term and long-term energy conservation plans. The commission and ministry have further collaborated on a program designed to build a water efficient society and on plans for water reuse in mineral pits and for seawater use. The two authorities jointly developed a broad program for developing a circular economy across China, as well as plans that apply specifically to major industries and fields such as iron and steel, nonferrous metals, coal, electric power, chemical production, construction materials, sugar, renewable resource distribution, and metal recycling and reuse. Six Chinese government authorities including the NDRC and the former SEPA jointly issued the *Notice on Conducting a Pilot Program for Developing a Circular Economy* (FGHZ [2005] No. 2199) in October 2005 with approval from the State Council. The first batch of pilot circular economy projects was launched for certain important industries, fields, industrial parks, cities, and provinces. The authorities launched the second batch of circular economy projects focused around energy conservation and emission reduction in major enterprises, industrial parks, cities, and provinces and in fields such as comprehensive resource use with the approval from the State Council in December 2007.[8] Local governments are pushing circular economy relevant laws, regulations, and standards at all levels. Seventeen municipalities and provinces including Beijing, Shanghai, Shandong, Chongqing, and Shenzhen have presented local regulation or government decision on circular economy development. The *Circular Economy Promotion Law of the*

---

[6] Huang [6].

[7] Ma [7].

[8] Zhou [8].

People's Republic of China was passed at the 4th Session of the 11th NPC Standing Committee on August 29th, 2008 and went into force on January 1, 2009.

The State Council approved implementation of the Gansu Provincial Program for Developing a Circular Economy on December 24, 2009, initiating China's first ever state-level circular economy demonstration zone in Gansu Province. The circular economy demonstration zone in this program will be built in 2015, creating seven circular economic centers across Gansu Province. The State Council approved the Qinghai Provincial Program for Building the Qaidam Circular Economy Test Zone co-developed by the NDRC and the Qinghai provincial government on March 15, 2010, which aims to build a state-level circular economy demonstration zone by the end of 2020. These first two regional circular economy development programs approved by the State Council highlight the state's strategic deployment and intention to adjust the economic structure and development pattern by promoting a circular economy. Four distinctive circular economy parks and six highly correlated and complementary industry chains have been planned for the Qaidam test zone to facilitate coordination between development rate and intensity with environmental protection. A commercial minerals prospecting system and a technology innovation and industrialization information service platform will be created in this test zone, in addition to institutional innovation such as the extended producer responsibility (EPR) program, pollutant discharge and emissions permit trading, compulsory cleaner production auditing, as well as circular economy statistics collection and assessment. The NDRC approved a state-level comprehensive economic reform test zone focused on resources in Shanxi Province on December 13, 2010 with consent from the State Council. This program's objectives include: accelerating industry mix optimization, strategic adjustments to the economic structure, technological advancement, and innovation by implementing serious reforms; furthering a resource-efficient environmentally friendly society; unifying plans for urban and rural development; and ensuring and improving people's livelihood. Chapter 23 of the *Outline of the 12th Five-Year Plan for National Economic and Social Development*, Great Efforts toward Developing a Circular Economy, expressed plans for circular economy development. "In promoting the development of a circular economy, we must adhere to reduction, reuse, and recovery, emphasizing reduction. In order to increase resource productivity, we should promote circular economy development in production, distribution, and consumption, and work faster toward building a framework for the circular utilization of resources that covers society as a whole." This document requires corresponding authorities to promote a circular production model in all sectors, improve resource recycling systems, and encourage green consumption. In the 12th Five-Year Plan period, the Chinese government authorities must enhance policy and technical support, including guidance on planning and on tax and financial incentives, to improve relevant laws, regulations, and standards and to implement the EPR program. The government authorities must also create circular economy technology and product catalogs, develop recycled product labeling, and improve the circular economy statistics collection and assessment system. Authorities should encourage the development and application of technologies for at source pollution reduction, waste recycling, remanufacturing, zero emissions, and cross-industry linkage and promote typical circular economy models. The government authorities must further implementation of state-level circular economy demonstration projects, including implementation of the "Ten, Hundred, and Thousand" demonstration program and promotion of nationwide pilot circular economy development areas such as the circular economy demonstration zones in Gansu Province and Qaidam, Qinghai Province, as well as the comprehensive economic reform test zone focused on resources in Shanxi Province.

1.2.4.4 Building a Resource-Efficient and Environmentally Friendly Society

The 5th Session of the 16th CPC Central Committee made the important decision of creating a resource-efficient and environmentally friendly society. Saving resources includes finding efficient ways to use land, building circular economy demonstration zones, and progressing resource pricing reform. Environmentally friendly efforts include establishing MFOZs, developing assessment indicators and policies for payment of ecosystem services (PES) and environmental restrictions, and improving the pollution discharge and emissions permit trading mechanism. The report to the 17th CPC National Congress emphasized that "adhering to the basic state policy of resource conservation and the environment is vital to the immediate interests of the people and the survival and development of the Chinese nation. We must give prominence to building a resource-efficient and environmentally friendly society in our strategy for industrialization and modernization and push every organization and family to act accordingly. We will improve laws and policies to promote energy, resource, ecological, and environmental conservation, and to advance formation of systems and mechanisms for sustainable development. We will implement a responsibility system for conserving energy and reducing emissions. We will develop and advance technologies appropriate for conserving, substituting, and recycling energy and resources and for controlling pollution, developing clean and renewable energy sources, protecting land and water resources, and establishing a scientific, rational system for a more efficient utilization of energy and resources." The NDRC green-lighted the Wuhan City Circle and the Changsha–Zhuzhou–Xiangtan City Cluster in December 2007 as two state-level comprehensive reform test zones, granting them permission to craft and test innovative policies in order to promote the society envisioned.

The cities of Changsha, Zhuzhou, and Xiangtan have replaced traditional size expansion models with a new concept that involves compact spatial arrangements, natural ecosystems that spatially separate the three cities from each other, and an eco-friendly and efficient cluster pattern for economic development. The core area was structured with a focus on Changsha and preservation of large green spaces. The metropolitan area forms a distinctive city cluster with close internal ties. Urban ecosystems and environmental protection are highlighted in the development of the Changsha–Zhuzhou–Xiangtan City Cluster in accordance with unified urban and rural development planning and a harmonious relationship between humans and nature, developing an indicator set geared toward assessing the environment for high quality and livability. The Wuhan City Circle focuses on testing a resource-efficient, environmentally friendly society built upon the innovation of nine basic systems and mechanisms. The pilot projects that seek creation of a resource-efficient and environmentally friendly society have played a significant role in finding effective and efficient strategies for creating primary eco-city groups or clusters in central China.

1.2.4.5 Building Districts and Cities as Pilot Areas for an Eco-civilization

Chinese government authorities such as the MEP have released indicators for assessing cleaner production in 24 industries such as coal-fueled power generation, phosphate fertilizer production, and lead and zinc smelting and have produced 25 industry standards for cleaner production, complementing development on the circular economy. The authorities have additionally issued three standards regarding green industrial parks, including the Guide for Pilot Eco-industrial Parks Planning (Pending), the Standard for Integrated Eco-industrial Parks (Pending), and the Standard for Sector Specific Eco-industrial Parks (Pending,) and have presented seven standards for controlling pollution by enterprises involved in waste recycling. Additional efforts by these government authorities include supporting laws and regulations relevant to a circular economy, promoting standards for resource recovery and reuse, and furthering labeling standards for energy and water efficient products and recycled products. The Chinese government is aggressively promoting the development of garden cities, the creations of cities that serve as environmental leader, and pilot eco-friendly districts, cities, and provinces in each region alongside its work on sector specific eco-friendly areas. The former Ministry of Construction (MOC) issued the *Implementation Scheme for Building State-Level Garden Cities* and the *Standard for State-Level Garden Cities* in May 2000, which cover urban landscaping organization, management, planning, design, and development, as well as ecosystem preservation and municipal infrastructure construction. They offer assessment indicators depending on the size and characteristics of the region and the city.[9] The *Ninth Five-Year Plan on Environmental Protection and Long-Range Objectives for the Year 2010 proposes construction of a number of clean and beautiful eco-cities with strong economic growth in order to promote environmental protection and sustainable development.* The former SEPA developed the *Assessment Indicators for State-Level Environmentally Leading Cities (Trial)* in 1997, before issuing the *Regulations on the Assessment and Acceptance of Beautiful Townships across China (Trial)* in July 2002. The former SEPA launched an initiative intended to build pilot eco-friendly areas in 1996 to promote eco-cities. It developed the *Outline of the Plan for Building Pilot Eco-friendly Areas across China (1996–2050)* to accomplish this objective. Additional efforts by the former SEPA included: proposing construction of eco-friendly districts, cities, and provinces in 1999;

[9]Zhang et al. [9].

developing the *Indicators for Eco-friendly Districts, Cities, and Provinces (Trial)* in 2003; issuing the *Indicators for Eco-friendly Districts, Cities, and Provinces (Revision)* on December 26, 2007; and issuing the *Scheme for Assessing the Building of State-Level Eco-friendly Districts and Cities* and the *Guiding Opinion of the State Environmental Protection Administration on Enhancing the Building of Pilot Eco-friendly Areas*. Construction of pilot eco-friendly areas has given much attention to green GDP practices. A joint research team set up by the National Bureau of Statistics (NBS), the former SEPA, the NDRC, and the State Forestry Administration (SFA) created the "Research on the green GDP accounting system" project in 2004. The team conducted in-depth research on green accounting practices for the national economy as they relate to basic frameworks, accounting methods, and theoretical foundations. The former SEPA and the NBS launched a pilot green GDP program intended for environmental accounting and for reporting on economic losses caused by pollution in ten municipalities and provinces including Beijing, Tianjin, Hebei, Liaoning, Zhejiang, Anhui, Guangdong, Hainan, Chongqing, and Sichuan in early 2005. The resulting "China Green National Accounting Study Report 2004" was published in 2006. This was China's first ever GDP accounting research report that adjusted for environmental pollution and represented the results of green GDP accounting efforts in the country over a determined period of time.[10] A Beijing green GDP accounting report was reviewed and accepted by the former SEPA and the NBS in November 2006, making Beijing the first area in China to complete the pilot program.

1.3 Eco-civilization: Building Eco-cities

1.3.1 What Is an Eco-city?

Eco-cities emerge from the human evolution toward the eco-civilization. The cities are built following theories derived from the eco-civilization. The eco-city concept derives from an in-depth review of industrial civilization-generated crises, such as resource depletion, environmental pollution, and ecosystem degradation. Theories and practices pertaining to eco-cities are collected from theoretical research across disciplines including urbanism, economics, and ecology.

1.3.1.1 The Eco-city Concept

The eco-city concept is first found in UNESCO's Man and the Biosphere (MAB) program launched in 1971. The concept has since evolved alongside social and technological development. An eco-city is a human settlement that is based on the

[10] Zhou and Huang [10].

sustainability of society, economy, human population, resources, and environment, and that is planned and designed with ecological principles, ensuring a harmonious society, efficient economy, and preserved natural ecosystems. An eco-city is a symbiotic structure in which nature, the city, and people form an organic whole. Eco-cities seek to form symbiotic structures with harmony among the city, people, and nature following eco-humanist theory. Eco-humanism aims to change human beings' economic and technological rationality mechanism and combine rational and value-based judgments by establishing ethics based on ecology and life. Eco-humanism relies on a tension between anthropocentric and non-anthropocentric world views but transcends the dichotomy. Eco-humanism considers the interests of non-human beings and all of the natural environment but still grants highest priority to human beings' common, long-term, shared interests. The philosophy hand picks traditional environmental ethics, retaining the central role of humanity present in anthropocentrism but rejects the anthropocentric master-and-servant relationship with nature, supporting the eco-centric belief that all species are equal and that humans and nature form an organic whole. Eco-humanism emphasizes the importance of ecosystems and the central role of humanity at the same time. Human beings hold a special position at the center of the harmonious whole of humanity and nature. The comprehensive development of humanity is the ultimate goal of all actions and measures so long as harmony between humanity and nature is maintained. Eco-humanism maintains the fundamental interests of humanity at its core. Thus, the philosophy gears ideas around humans and proposes methods that comprehensively meet human needs.[11] Ecological philosophy regards eco-cities as the achievement of a harmonious development of humanity and nature, presenting the epitome of resource-efficient and environmentally friendly societies. Eco-economics regards eco-cities as an efficient economic development method using circular economy and green industries. Eco-sociology regards eco-cities as ecosystems that feature coordinated societal, economic, human population, resource, and environmental development. Ecological urbanism regards eco-cities as structurally reasonable, functionally stable, complex ecosystems that integrate society, economy, and nature, where the flows of materials, energy, and information can be efficiently used. Urban planning regards eco-cities as human settlements with reasonable spatial structures, resilient infrastructure, popular eco-buildings, pleasant urban landscapes, and an optimized combination of built and natural environments. Eco-cities are geographical and spatial complexes in which urban and rural areas integrate with each other to promote green development.

1.3.1.2 Eco-city Characteristics

Eco-cities are complex ecosystems with a high degree of harmony uniting society, economy, culture, and nature. Internal material, energy, and information flows constitute a coordinated, symbiotic network with interlinked elements. Eco-cities

[11] Li [11].

incorporate material recycling, use energy efficiently, adjust according to information loops, maximizing economic benefit, as well as harmonize society and promote human and nature symbiosis. Eco-cities are symbiotic structures where nature, the city, and people form an organic whole. The primary characteristics differentiating them from traditional cities include:

A. Harmony. Eco-cities emphasize harmony between humanity and nature, the integration of nature with the city, and more importantly, harmony between people.
B. Efficiency. Eco-cities abandon modern city operations that indulge in high-energy consumption and create waste, maximizing the utilization of all physical and human resources. Both substances and energy are used at multiple levels and wastes are recycled. There is symbiosis across all sectors.
C. Sustainability. Eco-cities are built sustainably. Instead of engaging in what amounts to robbery to achieve temporary urban prosperity, eco-cities distribute resources accounting for space and time limitations, equitably satisfying the needs of the current and future generations in terms of development and environment. Healthy, sustainable, and coordinated development is ensured.
D. Holism. Instead of focusing solely on a beautiful environment or on personal prosperity, eco-cities consider the overall interests of the society, the economy, and the environment. They value coordination between economic development and the environment, but more importantly, a higher quality of life. Eco-cities seek development with harmony.
E. Decentralization. Eco-cities are regional in nature, combining urban and rural areas. Eco-cities correlate with each other and restrict each other to create a regional balance. Balanced and coordinated eco-cities require balanced and coordinated regions.

1.3.2 Eco-City Theoretical Foundations

Theoretical research on eco-cities has progressed steadily and has branched into many multidisciplinary directions. The theoretical foundations are analyzed in terms of ecology, humanities, circular economic theory, sustainable development, and the three form production theory.

1.3.2.1 Ecology

German biologist Ernst Haeckel coined the term "ecology" in 1869. He defined ecology as the scientific study of interactions among organisms and their environment. The *Encyclopedia Americana* defines ecology as a branch of biology that studies how plants and animals exist in nature and rely on one another. Ecology applied to human society is the scientific study of how human beings exist in natural and social environments and how they interact with these environments. The global environmental crisis has been worsening since the 1970s, posing unprecedented challenges to urban development. Ecological principles and methods have been applied in many levels of urban planning and development to address these challenges, leading to urban ecology. Urban ecology studies living organism relationships in an urban environment with an aim to adjust urban residents' relationship with the environment by introducing subsystem arrangements to the urban ecosystem. Urban ecology is also helpful for maintaining a balanced urban ecosystem and achieving harmonious, efficient, and sustainable urban development. The growing ecological, particularly urban ecological, studies have laid a firm theoretical foundation for eco-cities.

The ecological perspective views a city as an artificial ecosystem. An urban ecosystem is a human settlement or social, economic, and natural complex that connects human beings, resources, and the environment (including natural, social, and economic environments) within a certain area through various ecological, social, and economic networks. This peculiar artificial ecosystem is established as human beings change and adapt the natural environment. A city is a populous place with dense construction and many factories, shops, government organizations, schools, etc. A city is also a place where energy and resources are consumed, products and wastes are produced, and goods, wealth, and human resources are concentrated in large quantities. An urban ecosystem is similar to any other artificial ecosystem from the ecological perspective – a large, many layered, multifunctional system exists through natural and artificial component interactions and that incorporates human intervention. An urban ecosystem is a geographic complex oriented around humans that includes all factors within an urban space. An urban ecosystem is a typical complex ecosystem with economic, social, and natural subsystems. The natural sub-ecosystem mainly consists of biological and physical structures and includes organic populations such as plants, animals, and microbes, the natural environment that is comprised of urban climate, geology, topography, hydrology, land resources, etc., as well as the built environment. The social sub-ecosystem covers the urban population and typically has high population density and high household consumption, taking into account the systematic development of urban population structure and size, residential conditions, transportation, education, lifestyles, public organizations, etc. The economic sub-ecosystem centers on urban production and incorporates the conversion and flow of physical resources and energy. This subsystem is comprised of industry, agriculture, construction, transportation, trading, finance, scientific research, and education, along with other sectors. The natural, economic, and social subsystems of an urban ecosystem form an organic whole with close ties and interactions among each other. An urban ecosystem is a complex where natural ecosystems overlap and interact with the social and economic systems.

1.3.2.2 Humanism

Cities appeared and prospered as a result of human development. The original meaning of the term "city" is a civilized environment where people live. "For the city should be an organ of love; and the best economy of cities is the care and culture of men," wrote Lewis Mumford in *The City in History*.[12] This poetic expression embodies the human-oriented eco-city. Humanism's core values are attention to human needs and values, respect for personal choice, and the promotion of comprehensive human development. Respecting people's physical and mental needs, working toward a livable environment, and providing humanistic care are necessary to build an eco-city with a humanistic approach. It is necessary to attract citizens' participation, to emphasize public services, and to call for emotional identification. These traits permit people to lead safe and healthy lives filled with comfort, freedom, and harmony. Humanist theories put forth the following four guidelines for creation of eco-cities:

1. Creating livable spaces. Livability measures against the idealized social city. A convenient, comfortable, and safe urban environment suitable for all residents to live, study, and work is considered the most livable city. Livable spaces should be available across the city regardless of its size, offering specialized designs that embody humanistic care in abundance. A city becomes a place for people to interact with each only when certain elements and images satisfy their basic needs and match their code of conduct. A successful urban design shows an insight into how people perceive their environment, incorporating high aesthetics, thereby creating a pleasant, comfortable, and convenient environment, according to Kevin Lynch.[13] How can a livable environment be created in the city? Based on Agenda 21's requirements for building livable environments, building livable cities should include: (a) providing sufficient and diversified residential buildings for all city dwellers; (b) improving infrastructure including water and power supply, heating, drainage, sanitation, and garbage treatment facilities in order to provide citizens with reliable services; (c) providing sustainable energy and transportation systems to allow citizens to live and travel easily and conveniently; (d) providing diversified spaces for recreational, cultural, and sporting activities to bolster a sense of belonging and leisure opportunities; and (e) promoting sustainable planning and land use, construction, and environmental management to instill order, reason, and energy into spatial arrangements throughout the city.
2. Humanistic care in an eco-city. Cities are a product of human civilizations and carriers of human culture. Jane Jacobs thought cities primary function and meaning was to allow for the growth of humanity. Urban spaces should be more than a collection of public works or a product of technical efficiency, providing space for the growth of humanity and interactions between people, pumping lifeblood

[12] Mumford [12].

[13] Kevin [13].

into the urban environment.[14] Urban development should favor the interests of the largest possible number of ordinary people, adhere to the population's spiritual world, and highlight basic quality improvements. A positive living environment is a basic physical condition for everyone's life. A person can live with dignity only if this basic condition is satisfied. Among the basic objectives of building a moderately prosperous and harmonious society in China is providing an opportunity for most people to live and work contently. How should urban development balance improving what is already good and providing assistance to those in need in light of limited resources? Cities belong to the people and should express the general consensus rather than create social hierarchy, privilege, or stratification. Cities must be built with the general public's needs in mind, enabling the public to functionally satisfy its needs. An urban space is less a landscape than a service seeking to be safe, suitable, and convenient. Urban beauty is desirable only if it increases convenience. Bolstering urban beauty at the cost of functionality is too burdensome for ordinary citizens to afford. Similarly, improving already successful areas is important in urban development, but it is more important to prioritize the interests of the general public, especially disadvantaged groups and people with low and middle incomes, focusing on timely assistance to those in need.

3. Public engagement in an eco-city. The organic and affectionate collective personality, power, and glory of the city integrate the entire community, shared with even the humblest dwellers, according to Lewis Mumford.[15] Citizens' participation in urban development planning, the combination of thoughts, the blending of cultures, and institutional innovation generated by the gathering people together, will doubtlessly become a huge driving force for growing healthy cities. The energetic and valuable creativity of city dwellers on matters of comfort, safety, and beautification of the spaces and environment where they live comes naturally from their hearts, allowing for significant impacts. Citizen participation in urban development planning differs from planning based on elite decisions, creating cities more likely to develop in a healthy, democratic, humanistic, and harmonious manner because city dwellers' feelings, understanding, identification, and creativity are reflected in the urban development plans. This result arises for a number of reasons. First, a city should be a physical product of expert analysis but should more importantly incorporate and grant attention to citizens' emotions and wishes. Second, different people and groups have different values, which can only be respected if varying sets of values are expressed in the urban development plan. Third, urban development planning is a public right and a very important public power. The common objectives of urban development can be best maintained, and the urban development plan will be most reasonable, authoritative, and feasible if citizens participate in urban development planning and the public shares planning power with the government. Fourth, citizens' enormous potential dwarfs the knowledge and power of urban planning experts.

[14] Yu [14].

[15] Mumford [15].

Public wisdom and a public free to exercise their rights are necessary to maintain a positive city as a collective home.

4. Community harmony in an eco-city. Vibrant urban life does not result from an ordered environmental structure planned by engineers. People will keep inventing ways to use spaces and adapt streets, parks, squares, and playgrounds to their needs. Interactive adaptation enlivens public spaces in the city, and any environmental plan that opposes adaptations abandons the humanity of the environment. A successful city demonstrates social harmony and citizen cooperation. A variety of existing community activities energize and unify the community, making it safer. The city is not only a space where people live but also a meaningful and emotional environment. People often associate a particular meaning or emotion with a particular space. The term "home" is not confined to an apartment or house but can refer to a street or neighborhood, or even an area or community in a city. Comfort, safety, and familiarity associated with an apartment or house are also possible in other surroundings. In recognition of the importance of belonging to a city, safe, clean, beautiful, caring, and harmonious communities should be created. A city mixes humans with nature on many dimensions. It should be a desirable space. People feel alienated, depressed, and nervous in an urban environment that exceeds physical and mental tolerance levels. A harmonious community requires a humane space full of warmth and friendliness. These spaces should be visible and accessible. A park that is accessible for any part of a community in less than a 5-min walk is an example of a humane space, because it can be integrated into daily life. These spaces minimize social inequality caused by differences in income, power, reputation, or other aspects. A city should provide a link between hearts and souls, eliminating estrangement. It should be diverse, where people of different ethnic groups, ages, and preferences interact with each other in a pleasant and harmonious manner. A city should be reliable, inspiring trust and communication, as part of a beautiful and harmonious community.

A city is a human creation. It is a built environment humming with human activity and representing human wishes, cultures, and dreams. A city is the collection of its citizens' lives. It pursues civilized behavior and etiquette to ensure a good life for people. Aristotle said that people come to the city to live and stay for a better life. This sentence incisively reveals the essence of the city – it is not a mere physical entity but the container of human life. The humanistic eco-city tells us that people are the owners of the city and demand that it adhere to human nature, sow the seeds of fraternity, harmonize interpersonal relationships, and promote ethics and humanity. Such a city can be a stage for achieving comprehensive personal development and a harmonious society.

1.3.2.3 Circular Economy

Circular economy is an economic model that complements sustainable development. At the core of a circular economy is the efficient use and recycling of resources and adherence to reduction, reuse, and recovery. A circular economy fundamentally alters the traditional economic model that exhibits exorbitant production, consumption, and waste by promoting low consumption and emissions through high efficiency. A circular economy integrates cleaner production methods with comprehensive waste reuse. The economic model is ecological due to its emphasis on conservation and efficient use of resources and focus on material circulation throughout society. The carrying capacity of natural ecosystems is considered in calculations of a circular economy, natural resources are conserved whenever and wherever possible, and material and energy flows are recycled. A circular economy production process must obey the 3Rs – reducing production inputs in the form of consumption of natural resources, reusing products whenever and wherever possible, and recycling waste to minimize total waste volume and convert as much waste as possible into new resources. Renewable energy options, such as solar and wind power, replace nonrenewable sources in production whenever and wherever possible, pushing production to rely on circulation in natural ecosystems. Advanced technology is incorporated into systems so that knowledge replaces physical materials input whenever and wherever possible to achieve harmony between the economy, society, and ecosystems. A circular economy is realized by emphasizing cleaner production and green manufacturing during conversion to an eco-city. Cleaner production is an emerging discipline focused on industrial production patterns. Cleaner production has evolved as a basic model for enterprises to achieve eco-friendly production to include minimization or elimination of waste and comprehensive waste reuse. The production model prevents pollution in an attempt to avoid harm to the environment in all aspects ranging from raw materials to finished products. Cleaner production demands clean raw material extraction, environmentally friendly energy sources, and innovative production technologies so that products will not harm human health or ecosystems. Green manufacturing, also known as environmentally conscious manufacturing (ECM), is a contemporary manufacturing system that incorporates a comprehensive consideration of resource use and environmental impact. ECM aims to minimize the environmental impact of a product, eliminate the product's potential harm to human health, and maximize lifecycle resource efficiency, spanning design, manufacturing, packaging, transport, use, and disposal. Green manufacturing pushes a green outlook for all steps involved in manufacturing a product from design to production and packaging.

1.3.2.4 Sustainable Development

The respected Club of Rome published its first ever research report titled "The Limits to Growth" in 1972. This report offered an insight into the significance of the environment as well as the inherent ties existing between natural resources and humans. It emphasized that Earth's resources are unable to support the further development of consumption-based industrial societies around the world and that human beings must promptly change how they use resources and the prevailing economic models, instead seeking new socioeconomic development strategies in order to accomplish a coordinated and sustainable development trajectory of resources, the economy, society, and ecosystems and eliminate the imminent threat to human survival posed by the impending resource crisis. The WCED submitted a report titled "Our Common Future," also known as the Brundtland Report, to the UN General Assembly in October 1987. This report first introduced the concept of "sustainable development" after exploring a series of major economic, social, and environmental issues facing human beings. Sustainable development is "development that meets the needs of the present without compromising the ability of future generations to meet their own needs." It is the harmonious development of the economy, society, resources, and environmental protection in a holistic system. Economic development goals should not be compromised, but the environment and natural resources such as air, freshwater, oceans, land, and forests should be well protected such that future generations will be able to enjoy and not be forced to live and work in more resource-stressed conditions. Sustainable development focuses on development and requires socioeconomic development with tight population control, improved quality of life, the environment protected, and resources used in a sustainable manner. Human beings and the Earth on which they rely for survival and development can be said to constitute an enormous and complex human–Earth system using a system theory perspective, a social–economic–natural ecosystem complex in which all subsystems correlate, interact, and restrict each other. Population, resources, and environmental and developmental issues must be understood and addressed as a whole before an integrated human–Earth system is established in order to effectively develop sustainably. Sustainable development represents optimal development in recognition of the human–Earth system. Environmental economists see sustainable development encompassing correct handling of natural resource use and of discharge and emission of waste, increased awareness of the environment's value, promotion of effective resource use, environmental pollution minimization, new approaches to using resources, renewable resource maximization, and coordinated and simultaneous economic, social, and environmental benefits. Sustainable development is the pursuit of sustainable economic development that protects natural ecosystems on Earth as viewed by eco-economics. Economic development and ecosystem conservation should be achieved in a unified manner, creating unified economic, social, and ecological benefits. The sustainable use of resources and preserved ecosystems characterize sustainable development in this perspective. Sustainable development answers social ethical concerns by promoting the fair distribution of resources on Earth, satisfying current socioeconomic development needs without compromising long-term needs and considering the interests of the current generation without damaging the interests of future generations. Sustainable development asserts that humanity can be in harmony with nature. Human beings should raise sustainable development awareness and capacity. For balance, human population must be restricted to sustainable limits, and the pressure on resources and the environment must be reduced to achieve harmony between humanity and nature.

1.3.2.5 Three Production Classes

The world can be split into people and the environment, which come together through a complex system of human societies and the natural environment. This global system sees matter flow through three fundamental and interlinking production classes: material, human, and environmental. The relationships between the production classes have endured a long evolutionary process as human societies and knowledge have grown. Material, human, and environmental productions have been rendered three independent systems by industrial civilization. Correlating the three systems dominates the global system's operations.

Material production refers to the process in which human beings use technology to extract natural resources from the environment and receive regenerated materials from human production, before converting the resources into consumer goods. This process fulfills humanity's need for consumer goods but also generates waste that is disposed of in the environment. Human production refers to the process of human survival and population increase, including human reproduction and the consumption of material goods for the sake of survival. This process consumes consumer goods attained through material production and consumer resources attained through environmental production and provides human resources to support material and environmental production. Human production produces waste that is disposed of in the environment but also produces regenerated material that is sent for use in material production. Environmental production refers to the process by which the environment maintains and improves its natural structure and state using both natural and human forces. This process includes reception and absorption of pollutants (including waste from production and consumption) and generation of resources (including consumer and producer resources.) The theory behind the three production classes stipulates that these three production activities form a single, ring-shaped structure in which materials circulate. Obstacles blocking the flow of any material within this system disturb the harmonious flow and the sustainable development of this global production system. The extent to which humanity is in harmony with the environment is therefore dependent on how smoothly materials flow among the three production classes. Human beings in industrial civilizations are only aware of material and human production and their mutual relationship, overlooking environmental production. An incomplete understanding of the structure and operation of the global system has led to views, theories, methods, and technologies that induce chaos and render sustainable development impossible. Human beings are now already aware that many preexisting views, theories, methods, and technologies must be changed. Building environmentally friendly and resource-efficient societies and developing green industries and a circular economy are all part of necessary steps on the path to eco-cities and an eco-civilization.

The production classes theory highlights the main objectives behind developing an urban eco-civilization. Mapping the three production classes over the global system reveals how materials must flow smoothly within the system, and material input and output for all production must be balanced to achieve harmonious management. Human consumption and production waste must be disposed of in an environmentally friendly manner or converted into resources for the material production subsystem. The primary objective of developing an urban eco-civilization should be matching natural resource exploitation and waste generation with environmental productivity.

The three production classes theory identifies primary control targets in eco-city construction. The theory shows how environmental issues generally occur where different production systems interface and overlap. Building eco-cities should focus on the behavior and activities of human societies existent on various interfaces. State parameters relevant to production must be controlled to within certain ranges to harmonize material flows among the three production classes.

The three production classes theory presents a foundation for an urban eco-civilization. It reveals the essence of the human–environment relationship and of the interactions between human societies and the natural environment at the material level. This theory will help human societies change the current lifestyles by bringing attention to issues that threaten human health and the environment inhabited by humans. Humanity must harmonize its relationship with the environment to achieve sustainable development by correctly managing social behavior and organizing efforts in the management of urban eco-civilization creation. The theory proposes that humanity must plan eco-cities by managing social behavior and lead urban development toward an eco-civilization.

1.3.3 Eco-cities: Planning Principles, Objectives, and Standards

Planning Principles Eco-city planning principles date back to amenity planning concepts developed in the United Kingdom in the 1950s and 1960s. The Civil Amenities Act passed in the United Kingdom in 1967. Livability in the context of urban development focuses on: improving environmental health by eliminating crowded and noisy environments where diseases are rampant, creating harmonious and beautiful environments to provide people with comfort and pleasure as well as satisfy their spiritual needs, protecting the natural environment and building cities upon preserved ecosystems and harmony with nature, and protecting historic and cultural elements and increasing the environment's cultural significance. Five eco-city planning principles were proposed under the MAB program in 1984, including the ecosystem conservation strategy, eco-infrastructure, living standards for residents, protecting artifacts of cultural and historical importance, and integrating cities with nature. These principles summarize the primary eco-city planning ideas.

US ecologist Richard Register offered ten recommendations for building eco-cities in 1990, including: increasing ecological awareness among the public; dedication to researching technologies and measures that improve material and energy circulation inside and outside cities and reduce nonrenewable resource consumption while more effectively conserving and using renewable resources; establishing management departments for eco-city creation and improving the current eco-city creation management system; ecologically rebuilding cities and creating diversified spaces for residents; creating and restoring wildlife habitats; adjusting and improving the urban eco-economic structure; enhancing the ecological restoration of older cities and of unused land in cities; establishing a reliable public transportation system; canceling vehicle purchase subsidies; and creating policies that encourage individuals and businesses to participate in eco-city creation. These ten recommendations offer a comprehensive view of the most relevant issues and trends in eco-city creation in the West.

Ecocity 5 was held in the city of Shenzhen, China, in August, 2002. The *Shenzhen Declaration on Eco-city Development* was passed at this conference. This declaration proposed five eco-city planning principles. The first principle is environmental safety. All city dwellers should have access to clean air, safe water, food, housing, job opportunities, municipal services, and disaster protection. The second principle is environmental health. Urine, feces, sewage, and garbage should be treated and reused by environmentally friendly means that are efficient and inexpensive. The third principle is eco-friendly industrial development. Industries should undertake eco-friendly transformations and to enhance the resource reuse, product lifecycle design, renewable energy development, and green and efficient transportation technology. Residents' daily needs should be satisfied, while resources and the environment are protected. The fourth principle is the combination of environmental protection and landscaping. Convenient urban transportation should be provided to all city dwellers by integrating the built environment, open spaces (e.g., parks and squares), and hubs such as streets, bridges, and natural elements (e.g., watercourses and city skylines) to conserve energy and resources and minimize traffic accidents and air pollution. Damage to the aquatic environment should be prevented and the urban heat island (UHI) effect alleviated to lessen the impact on the worsening global environment. The fifth principle is the development of environmental awareness. People should be assisted and made aware of how they relate to nature and of their environmental responsibility. Consumer behavior guidance should be provided to change the traditional consumption pattern and strengthen the autoregulation capacity, thereby maintaining a high quality in the urban ecosystem.

Objectives The first eco-city objective is to create sustainable and efficient economic development. The loosely controlled, expansion-based industry system should be replaced by an economic system that is efficient and conserves resources. Individuals and organizations should work to increase efficiency, conserve natural resources and energy, and reduce waste. Relevant government authorities should encourage information technology in industrial development to further alter economic development and increase high-tech industry presence, advanced manufacturing, modern services, and agriculture in the industry mix. Product production should be gradually replaced by function production by combining production, consumption, transportation, and recycling systematically, to create an interconnected web and horizontal control and communication between enterprises and their corollary departments, providing vertical optimization of product lifecycles and connecting sectors of industrial production, surrounding agricultural production, and social systems. A circular economy industry system should be created to improve resource efficiency and encourage industries that generate high economic and social benefits. Natural, human, and social resource potential should be used to its fullest extent. Sectoral resources should be centralized to optimize distribution. The second objective is civilization development. Human population size and resource availability must be kept in balance; public health and civilization developments are simultaneously necessary, along with human development opportunities. A diversified education system should be created for various levels to better educate the public. A social environment that assures equality, freedom, and civil behavior should be ensured. Continuous improvements to social security and service systems are necessary to ensure societal safety and stability. The third objective is ensuring a healthy environment. A healthy regional environment, maintained ecosystem services and functions, and ecosystem safety are all crucial. Ecosystem rehabilitation and restoration mechanisms and a unified open green space system are necessary for urban and rural areas. Ecosystem resource treatment, recovery, and reuse systems require continuous improvement if they are to provide city dwellers with a clean, healthy, and safe environment. Proper land use is another important factor. Optimizing spatial arrangements and functional zoning in urban and rural areas will help establish an eco-friendly spatial structure. Integrating built and natural environments will generate innovative physical structures, ecological functions, and aesthetic effects and also gear urban landscapes toward the needs to humans and the ecological surroundings. Environmental efforts focused on ecological engineering and ecosystem management and services, moving away from environmental engineering and management and landscaping, will achieve greater harmony between built and natural environments. The fourth objective is to bring about cultural and ecological harmony. Historical and cultural heritage require protection; local cultures including folklore require respect to maintain cultural diversity. Eco-civilization concepts can be disseminated into fields such as policy, institution, production, and life by introduction a new cultural system centered on ecology and advocating ecological ethical values. The consumption mix can be improved if more sustainable consumption is adopted. Consumption benefit increases, and green lifestyles adopted across society are also helpful in this pursuit.

Standards Eco-city development standards are established following social, natural, and economic ecosystems. Social ecosystem principles include human-oriented concepts, satisfaction of human beings' physical and spiritual needs, and a social environment that assures freedom, equality, fairness, and stability. Economic ecosystem principles include natural resource and energy conservation and efficiency, improved resource recovery and reuse, efficient resource use, and sustainable production, consumption, transportation, and residential area development. Natural ecosystem principles include prioritizing natural ecosystem considerations to maximize its protection and restricting human activities in consideration of the natural environment's carrying capacity while reducing their negative effects on the natural environment. Eco-city construction includes eco-infrastructure, eco-friendly living environments, eco-metabolic networks, and an ecological capacity.

Eco-infrastructure provides an urban eco-civilization with a support system. Eco-infrastructure includes urban drainage systems; regional energy supply, light, and heat dissipation systems; urban soil activity and soil percolation systems; urban ecosystem services and biodiversity networks; urban metabolic and venous circulatory systems; regional airflow fields; and the underlying ecological pattern. Eco-infrastructure construction aims to maintain the systems' structural and functional integrity and its ecological activities and also to enhance the water, soil, air, ecosystems, and mineral resources' carrying capacity. There are a number of indicators used to measure eco-infrastructure:

The percentage of water used for non-ecological purposes measures the ratio of water used for urban production and consumption and the multiyear average amount of water resources needed to maintain the local natural ecosystem's basic functions. This ratio is best kept below 35 %. The ratio of land used for ecosystem services measures the ratio between the total area of land used for agriculture, forestry, green spaces, wetlands, and nature reserves in urban areas and the total area of land used for urban development. Ideally the area of land used for ecosystem services is at least twice as large as the area of land used for urban development. Renewable energy use such as geothermal, solar, wind, and biomass energy should remain above 10 %. The percentage of the total area suffering from a strong UHI effect (a temperature difference exceeding 2°) should not go above 10 %. The ecological threshold and the percentage of local species should remain at least at 65 %. It is also desirable to increase landscape diversity and decrease disasters.

An eco-friendly living environment is a pivotal part of an urban eco-civilization. A city's living or industrial environment's sustainability depends on the structural and functional integrity of green spaces, wetlands, ground surfaces, main outfalls, mountain and river systems, and major roads in communities or industrial parks. Ecocity 8 attendees proposed developing a set of living environment standards for eco-cities in Istanbul, Turkey, in 2009. The following are ten recommendations based on global experience building eco-cities:

A. Compact spatial arrangements: building up from ground surfaces and into the ground. A particular emphasis is needed on street and underground development. Interactive residential areas where building has 6–10 levels are most efficient because fewer levels fail to optimize land use, while any more levels render buildings inefficient and return lower social and environmental benefits. The population density in a community should be no less than 10,000 people/km^2. Strengthen major arterial roads in the city: new urban areas and industrial parks serially arranged along light-rail lines and high-capacity, high speed arterial roads will provide public transportation between these sectors and central urban areas. Functional groups separated from each other by natural land such as green spaces, wetlands, farmland, and woodland will boost ecosystem services. An eco-friendly transportation network should extend to more than 80 % of the city's urban population. The standard requires that a bus ride from any station on a major artery road going directly to the downtown area not exceed 30 min. Dispersed micro green spaces: green spaces in an apartment complex should be extended as long as possible. A walk to the nearest large green space should take no more than 10 min.

B. Resilient wetland and green space ecosystems: permeable city surfaces that allow plants to grow through will increase resiliency. Rooftops and facades should be covered with plants. Sunken green spaces should double as wetlands, providing water supply and drainage. Urban areas should dedicate at least 30 m^2 per capita toward ecosystem services, including wetlands that measure at least 3 m^2 in area. Mixed zoning providing multiple functions alongside work and residence: residence, industry, commerce, administration, and ecosystem services should be combined with each other. At least one third of employees should be able to work near home. A bus ride between the home and the workplace should not take more than 30 min for most residents.
C. A convenient and eco-friendly public transportation system: at least 80 % of residents should travel by bus, light rail, or bicycle during rush hour. A walk from any location in the city to the nearest bus station should not take more than 10 min. Percentage of green buildings: A newly built community should be composed of at least 70 % green buildings. Green buildings consume 60 % less energy and emit 50 % less carbon dioxide than traditional buildings (15–30 % less fossil energy).
D. Highlighting eco-branding: iconic buildings, statues, and biological and cultural landscapes should be used to bring attention to the local ecosystems, both human and natural. Over 80 % of the population should be satisfied with eco-branding efforts.
E. Greenways for sightseeing and ecosystem services: continuous greenways unaffiliated with the bus and light-rail system should be available to citizens and visitors for sightseeing and to bolster ecosystem services. The greenways should include a network of cycle paths and sidewalks, recreational spaces, service facilities, green fences, and buffer corridors. At least 5 m^2 per capita should be dedicated to urban greenways. Greenways should cover and connect every community, village, and tourist destination within municipal boundaries.
F. Honest and friendly neighborhoods: residents should be able to feel camaraderie and security in their community, where a full range of recreation and sports facilities and spaces should be available. Residents should actively participate in community activities. At least two thirds of residents should be able to meet other residents in their community on a daily basis and communicate more broadly on a weekly basis.

The eco-metabolic network is the operating system for the urban eco-civilization. An urban eco-metabolic network organizes and adjusts itself to allow for a high-intensity flow of energy, materials, information, money, and people. The network is continuously consuming resources to revitalize, it engages in production, distribution, consumption, restoration, and control, and it follows a pattern from gestation, through development, prosperity, maturity, and decline to rejuvenation. The eco-economic efficiency and environmental impact of this network can be measured with the following indicators:

A. Urban ecological footprint: the total land area needed to maintain a city's basic consumption and absorb its waste. A city's ecological footprint can be reduced by increasing natural resource output per unit area, efficiently using existing resources, and altering the production and consumption patterns.
B. Urban ecosystem services: the ecosystems' ability to provide tangible or intangible natural products, environmental resources, and ecological public benefits needed to maintain the urban society's production, consumption, distribution, restoration, and control activities. These services are accounted for using spatial, temporal, equivalent, arrangement, and order measures. The ecological efficiency of industries measures how efficiently the industrial system's ecological resources satisfy the city's needs. This efficiency ratio checks the product and service outputs against resource and environmental inputs. A product's entire life cycle is assessed, and all stages from the extraction of natural resources, materials processing, production, transportation, and consumption to recycling are analyzed. This practice aims to find the right economic methods and policies that will increase the industrial system's ecological efficiency.
C. Eco-friendly material flow: the city's local food production or consumption should account for at least 50 % of the total food production or consumption, respectively. Sewage treatment and water conservation and recycling facilities should all be efficient. The average domestic water consumption per capita should be less than 100 liters/day. Environmental health measures should be widely used in urban and rural areas. A household compost pit on average 1.0 m^2 in area should be made in every community so that 70 % of garbage can be locally converted into resources.

Ecological capacity building provides the assurance system for the urban eco-civilization. Urban eco-civilization capacity building is a humanistic process that focuses on changing people's values, adjusting interpersonal relationships, regulating people's behavior, and improving comprehensive capacities. The entire urban environment can be improved by adjusting the behavior of all stakeholders and building capacity. These adjustments will contribute to a harmony between humanity and nature and between people themselves developing in urban civilization. Synergy and symbiosis throughout the region, prosperity of urban and rural areas collectively, institutional consolidation, harmony between human and nature, and a balanced society comprise goals that should be worked toward. The system should focus on balancing local and global interests, short-term and long-term benefits, growth rate and quality, as well as internal innovation and cooperation with the outside world. All sectors, organizations, and classes in the city are advised to compete, work symbiotically, revive withering areas, and develop themselves. The following indicators can be used to measure this progress:

A. The ecological cognition index: the index covers ecological knowledge (including ecological philosophy, science, engineering, aesthetics, and economics) of decision makers, entrepreneurs, technology experts, and ordinary people, ecological awareness (regarding global environmental changes, regional ecosystem services, environmental health of certain peoples, and ecological sustainability management), and the level of ecological concern (about subsistence, benefits, morality, beliefs, and the whole world).
B. The capacity to make an ecologically unified plan: the capacity of urban administrative departments to make unified plans for the region, urban and rural areas, harmony between humans and nature, as well as for the society and the economy.
C. Economic vitality: a combination of the carrying capacity and potential of natural resources in the heartland, the carrying capacity of ecosystems, and developing technology and human resources capacity, industry composition, and arrangement with regard to ecological impact, the percentage of researchers, developers, service staff, and trainers, economic strength, growth rate, diversity, and stability.
D. Public participation potential: a combination available mechanisms, systems, and platforms for public participation, public motivation to participate, knowledge and skills relevant to making important decisions and finding importance in outcomes, bottom-up approaches to public monitoring, and volunteer participation.

1.3.4 Metrics and a Dynamic Assessment Model

Eco-cities evolve over time. A dynamic eco-city assessment model centers on descriptions and assessments of an eco-city's current state. Established metrics are preferred for this description. Sequential relations are critical to properly describe issues because metrics are dynamic. Temporally sequential relationships are used to determine whether an eco-city is developing sustainably. Sustainability ranks high among eco-cities development measures. Assessing eco-cities' development requires objective metrics because development is otherwise highly relative and difficult to compare or accurately describe. Dynamic metrics use assessment criteria established depending upon the highest and lowest developmental levels of eco-cities studied during different periods. Only after assessment criteria are established can the current eco-city development be measured by calculating the difference between observed data and the standard values or by recording the frequency of a certain indicator exceeding some standard value.

1.3.4.1 Metrics

The UNCSD developed 134 social, economic, environmental, and institutional indicators (and later on simplified them into 58 indicators) under the DSR model. In China, the former SEPA issued *The Indicators of the Development of Eco-districts, Eco-cities, and Eco-provinces* in 2003. These 22 indicators cover economic development, environmental protection, and societal advancement. The former MOC issued the *National Standard for Eco-Cities and Garden Cities (Trial)* in 2005. This standard includes 19 indicators that cover urban ecosystems, the living environment, and infrastructure. Eco-city development can be broken down into indicators using a comprehensive assessment based on a complex ecosystem perspective. The comprehensive eco-city assessment system breaks down into eight Level 2 indicators using the eco-city assessment system design principles, including the ecosystem, the living environment, resource conservation, economic development, lifestyle, humanistic policy, the environment, social development, and social harmony.

The ecosystem indicator includes five Level 3 indicators – forest coverage, biodiversity, nature reserves composition of total area, soil erosion rate, and natural disaster frequency.

The living environment indicator includes 11 Level 3 indicators – the air quality index, the average area of public green space per capita, the water quality of rivers and lakes, average carbon dioxide emissions per capita, the percentage of drinking water that meets standards, the percentage of area where ambient noise levels meet standards, the percentage of garbage converted into harmless materials, the percentage of industrial waste water that is discharged in a manner that meet standards, the percentage of urban sewage treated in a centralized manner, urban beauty, and citizen satisfaction.

The resource conservation indicator includes five Level 3 indicators – reliance on resources and energy, food self-sufficiency, the average amount of water resources per capita, clean energy use, and the size and percentage of new (or renewable) energy.

The economic development indicator includes 15 Level 3 indicators –the percentage of GDP devoted to R&D, technology R&D capacity and the number of patents, the percentage of GDP derived from services, high-tech industries' growth rate, the cultural industry's growth rate, the percentage of GDP devoted to education, the percentage of GDP use for environmental expenditures, the percentage of agriculture practicing eco-agriculture, the size of the circular economy and its percentage of the overall economy, overall energy consumption per unit of GDP, water consumption per unit of GDP, carbon dioxide emissions per unit of GDP, sulfur dioxide emissions per unit of GDP, the composite reuse of industrial solid waste, and industrial waste water recycling.

The lifestyle indicator includes ten Level 3 indicators – citizen opinions, citizen-led green consumption, average energy consumption per capita, average domestic water consumption per capita, the percentage of green buildings, the average road surface area per capita, the percentage of green vehicles, the percentage of green trips, the percentage of citizens with environmental complaints, and the percentage of citizens volunteering for the environment.

The humanistic policy environment indicator includes seven Level 3 indicators – urban culture, existing government policies and regulations and their implementation, the percentage of government procurement from green sources, the public security index (or the frequency of criminal cases), the dissemination of environmental knowledge, laws, and regulations, corporate social responsibility (including donations toward environmental protection), and the status of regional and international cooperation (based on expert review).

The social development indicator includes nine Level 3 indicators – GDP per capita, the average number of years in school, life expectancy, the Gini coefficient, the Engel coefficient, the employment rate, urban–rural integration (or urban–rural income ratio), social insurance coverage, and medical insurance coverage.

The social harmony indicator includes three Level 3 indicators – social harmony, citizens' sense of happiness, and harmony between humanity and ecosystems, the environment, resources, the economy, and society.

These eight Level 2 indicators and 65 Level 3 indicators constitute the entire indicator set. Weighting and summing all of these indicators provides a comprehensive assessment of an eco-city for a given period of time. Open international standards are now available for reference for some of these indicators. Standards based on observational experience are required for standards without an open international reference. The highest and lowest values of any indicator can be taken from a given period and then weighted and averaged to establish an assessment standard for this indicator relevant to the next period. These weights are determined by using the optimal solution to the given problem with recognition of certain objective constraints.

1.3.4.2 Dynamic Assessment Model

The eco-city development standard is a dynamic concept. The development of eco-cities requires a set of development standards and indicators that reflect eco-city characteristics. Eco-city development standards and assessment methodology for China have been established using a new perspective that includes cones and semi-ordered space. A dynamic perspective divides eco-cities into three classes based on development performance. Class 1 includes cities whose eco-city development performance is growing over a certain period and that meet eco-city development requirements. These cities are referred to as healthy eco-cities. Class 2 includes cities whose eco-city development performance is growing over a certain period but that do not yet meet eco-city development requirements. These cities are referred to as sustainable eco-cities. Class 3 includes cities whose eco-city development performance shows both positive and negative trends over a certain period.

The dynamic perspective divides the development of Chinese eco-cities mathematically into early, intermediate, and advanced stages based on eco-city development performance. In the early stage cities of all three classes coexist; in the intermediate stage all cities have advanced to Classes 1 and 2; and in the advanced stage only Class 1 cities remain. The core of this model requires a set of indicators and development standards, because eco-city classification and the three stages of eco-city development in China are both determined by the cities' eco-city development performance. These indicators are divided into two categories – objective indicators, such as those described above, and subjective indicators. The subset of objective indicators are used to establish the lowest and highest levels of eco-city development across China during different periods, and a mathematical model can then be applied to establish development standards for different periods. Eco-cities are grouped into three classes according to the subset of objective indicators and the development standards. The three stages of eco-city development in China are thus conceptualized.

The assessment standard for eco-city development in China should depend on the current situation and on regional characteristics. This standard should rely on ecological principles and sustainable development theory. Harmony between humanity and nature must never lose importance, and environmental protection and resource conservation must also remain prominent. A balance between economic development, social advances, and environmental protection must be achieved and maintained. The size of an urban population should remain controlled, and urban and rural development plans must be unified. The eco-city development indicator set must be developed based on the lowest and highest development levels of different types of cities.

References

1. Yang T (2007) Environmental ethics: global discourse, Chinese perspective. Chongqing Publishing House, Chongqing, pp 1–4
2. Yang T (2009) The modern value of the traditional Chinese culture from an eco-civilization perspective. Environ Educ 5:25–29
3. Gang Z (2010) Regional development based on eco-civilization: new patterns and paths. J Yunnan Normal Univ (Human Soc Sci) 5
4. Qin D, Zhang K, Niu W (2002) Population, resources, environment, and sustainable development in China. Xinhua Publishing House, Beijing, pp 1–9
5. Hu J (2007) Hold high the great banner of socialism with Chinese characteristics and strive for new victories in building a moderately prosperous society in all respects. A collection of documents regarding the 17th CPC National Congress, the People's Publishing House, pp 13, 14 and 18–20
6. Huang X (2004) Circular economy: industrial pattern and policy system. Nanjing University Press, Beijing, p 18
7. Ma S (2007) On circular economy indicator and support systems, China Economic Press, Beijing, pp 69–103
8. Zhou H (2008) The latest developments of the circular economy in China. Recycl Resour Circ Econ 8:4–7 and 10
9. Zhang K et al. (2003) Urban assessment and indicators. Chemical Industry Press, Beijing, pp 6–15

10. Zhou Y, Huang G (2009) The development of Green GDP and its applications in China, Environ Sci Surv 2:11–24
11. Li X (2011) On the harmony and symbiosis between humanity and nature: all from Eco-civilization. China Zhigong Publishing House, Beijing, p 76
12. Mumford L (2005) The city in history: its origins, its transformations, and its prospects. China Social Sciences Press, Beijing, p 8
13. Kevin L (2001) The image of the City. Huaxia Publishing House, Beijing, p 34
14. Yu H (2005) Urban theory: classic and contemporary readings. Fudan University Press, Beijing, p 137
15. Mumford L (2005) The city in history: its origins, its transformations, and its prospects. China Social Sciences Press, Beijing, p 324

Chapter 2
Report on China's Eco-City Development

Weiping Sun, Juke Liu and Dingjun Wang

China has entered a critical growth stage where the percentage of population living in urban areas is rising rapidly. Strategic considerations during rapid urbanization include reviewing the current urban development pattern, abandoning traditional approaches to urban development, seeking sustainable eco-city development approaches, and improving urban development coordination with environmental protection and resource conservation. The report to the 17th CPC National Congress clearly expressed that China will "promote a conservation culture by ensuring that industry structures, growth patterns, and consumption habits become efficient in use of energy and resources and are environmentally friendly. We will implement a large-scale circular economy and considerably increase the proportion of renewable energy sources out of total energy consumption. Major pollutant discharges will be brought under effective control and the resulting ecological and environmental quality improvements will be notable. Conservation awareness will be firmly established throughout society." This statement adds new components to the traditional definitions of civilization and development, opens the door to a new era of sustainable development in China, and specifies new tasks and requirements for urban development and management. Urban development is changing its focus onto a better quality life, away from simple economic growth. Eco-cities have become a widely accepted model to develop toward the field of urban development in China.

W. Sun (✉)
Center for Social Development, Chinese Academy of Social Sciences, Beijing, China
e-mail: swp1966@126.com

J. Liu • D. Wang
Lanzhou City University, Lanzhou, China
e-mail: ljk@lzcu.edu.cn; 519542428@qq.com

J. Li, T. Yang (eds.), *China's Eco-city Construction*, Research Series on the Chinese Dream and China's Development Path,
DOI 10.1007/978-3-662-48153-0_2

2.1 The Current Status of Eco-City Development in China

2.1.1 *Eco-City Development in China*

China is working through the earliest eco-city development stage. Although the work has provided some experience and led to certain achievements, China faces problems and challenges in the continued development of eco-cities.

The Initial Eco-City Construction Wave Eco-city development in China has already surpassed several milestones. The city of Yichun in Jiangxi Province was the first Chinese city to announce its goal of developing into an eco-city in 1986. The local government drafted an eco-city development plan and launched a pilot program in 1987, pioneering efforts in China toward eco-city construction. Several scholars including Huang Guangyu developed an eco-city development plan for the city of Leshan in Sichuan Province in 1987. The Chinese government has been conducting pilot eco-civilization demonstration zone initiatives since 1995 such that today there are 389 state-level eco-civilization demonstration zones across China. The Hainan Provincial People's Congress promulgated the *Decision on Building an Eco-province* and approved the *Plan for Developing Hainan into an Eco-province Outline* in 1999, when the former SEPA authorized Hainan Province to join the national pilot eco-province development program. The former SEPA also authorized the city of Guiyang in Guizhou Province to join a national pilot eco-city development program in 2002 that was specifically focused on circular economy development. Guiyang has since provided significant contributions to national efforts that push cities rich in resources to develop eco-city structures. The Chinese and Singaporean governments jointly held the opening ceremony for a project known as the Sino-Singapore Tianjin Eco-city (SSTEC) in 2007, the world's first international eco-city development project. This project signaled the aggressive shift China decided to adopt in international cooperation for eco-city development. Eco-friendly reconstruction was an important objective in the efforts after the 2008 Sichuan earthquake. Some experts insisted that post-quake urban reconstruction focus on building small- and medium-sized cities that will be safe, comfortable, and environmentally friendly.

Another wave of eco-city development swept China since 2000, thanks to the implementation of the sustainable development strategy and increased awareness of the eco-civilization concept and the Scientific Outlook on Development. Jilin, Heilongjiang, Fujian, Zhejiang, Shandong, Anhui, Jiangsu, Hebei, Guangxi, Sichuan, and Liaoning provinces have announced their intentions to become eco-provinces. Beijing, Shanghai, Shenzhen, Suzhou, and Ningbo have likewise announced their intention to become eco-cities. As of February 2011, 230 of 287 Chinese cities at the prefecture level or higher, that is 80.1 % of these larger cities, had expressed their intentions and aims of developing into eco-cities, according to statistics from the Chinese Society for Urban Studies (CSUS) academic exchange department. Of the same 287 cities, 46.3 % or 133 cities had announced plans for becoming low-carbon cities. The combined result is that 90.2 % of cities, that is 259 of the 287, have announced plans to become eco-cities, low-carbon cities, or both.[1] Eco-city developed has moved to the top of the priority list for Chinese cities developing under the Scientific Outlook on Development and transforming their economic development pattern.

Eco-City Development Standards Multiple indicator sets exist to measure eco-city or sustainable development in China. These indicator sets are released by different government authorities, and each one focuses on a particular facet of the city, including the former SEPA indicators for assessing which Chinese cities serve as environmental protection role models in the 11th Five-Year Plan period; the former SEPA indicators for assessing and accrediting pilot eco-civilization demonstration zones across China; the former MOC national standard for garden cities; the former MOC *National Standard for Eco-cities and Garden Cities (Trial)*; the former MOC reference indicator set for the China Habitat Award; the CSUS scientific assessment standard for livable cities; the NDRC, NBS, and former SEPA circular economy indicator set; the SFA indicators measuring a National Forest City; the National Afforestation Commission (NAC) national inspection and scoring standard for cities that serve as landscaping role models; the Communist Party of China Central Guidance Commission for Building a Spiritual Civilization (CGCBSC) National Civilized City Assessment System (Trial); the National Patriotic Public Health Campaign Committee national clean city standard; the MOST primary indicators of socioeconomic and environmental conditions in the national pilot sustainable development zones; and the China National Tourism Administration (CNTA) national inspection standard for excellent tourism-oriented cities. *The Indicators of Eco-districts, Eco-cities, and Eco-provinces Development (Trial)* was released by the former SEPA on May 23, 2003 and includes 28 eco-city indicators that span economic development, environmental protection, and societal advances. A revised version of this standard, published on January 15, 2008, reduced the number of eco-city indicators to 19. According to the eco-city indicators' descriptions, the primary elements present in an eco-city include strong and balanced ecosystems; low environmental pollution; effective natural resource protection and use; a stable, reliable, and ecological safety assurance system; effective enforcement of environmental laws and regulations; faster socioeconomic development through a circular economy; harmony between people and nature as well as progress in green culture development; ability to create and maintain clean and beautiful urban and rural environments; and a higher quality of life in all respects. A Chinese municipality must satisfy the following requirements to be named an eco-city: a published eco-city development plan, approved and promulgated by the municipal people's congress for implementation; effective enforcement of local and national environmental laws and regulations; an independent environmental protection agency at the district or municipal level; factoring environmental protection in the CPC committee and district, city, and municipal governments' performance assessments of officials

[1] Li and Liu [1].

using an established assessment mechanism; ability to achieve energy conservation and emission reduction targets set by higher government departments; absence of severe environmental incidents over the previous 3 years; effective responses to the public's environmental complaints; no significant presence of invasive foreign species in local ecosystems; leading in the environmental quality index among municipalities in the same province; at least 80 % of all districts (or district-level cities) in the municipality must be recognized for meeting the national eco-district development requirements; and the central city is recognized as a role model in environmental protection based on assessment standards. The standard was significantly revised in its second draft. In the new standard, the ability to achieve energy conservation and emission reduction targets specified by the higher governmental departments is established as a basic eco-city requirement. "Environmental protection" was renamed as "ecological and environmental protection" to apply over a wider range. The indicator set was reduced to 19 by eliminating indicators deemed insignificant or too difficult to calculate. Only three societal advancement indicators remain in the new set. Some indicators were renamed and adjusted to more aptly guide eco-city construction. Water consumption per unit GDP was replaced by freshwater consumption per unit of industrial added value and the agricultural irrigation water use coefficient to more specifically reflect the pressure on water resources caused by economic development. The standard values of multiple indicators were adjusted to make eco-city development targets more consistent with relevant national policies. The standard value of energy consumption per unit of GDP was reduced to 0.09 tons of standard coal/1,000 yuan of GDP from the original 0.14 tons of standard coal/1,000 yuan of GDP. This stringent requirement aligns with the national policy regarding energy conservation and emission reduction. The indicators are now grouped into reference and restrictive indicators. The indicators are all restrictive except for those measuring the percentage of the tertiary sector in GDP and the three societal advancement indicators.

Different Types of Eco-Cities Various types of eco-cities can be found across China. The first type includes some newly built satellite towns that are classified as eco-cities, such as the first batch of pilot eco-cities, which include the SSTEC, Caofeidian International Eco-city, Wuxi Eco-city, Zhuzhou Eco-city, Fengxian Eco-city in Shanghai, as well as the Guangming and Pingshan New Areas in Shenzhen. China has built a dozen of these eco-cities, which already in their planning stages emphasize energy conservation, emission reduction, environmental protection, biodiversity, improvements to the living environment, and harmony between people and nature. The second type of eco-cities is built upon existing cities. The Chinese government has developed a series of standards for this process including the national standards for garbage classification and recycling, water recycling, green transportation, green buildings, low-carbon industries, and renewable energy. Existing Chinese cities may evolve into eco-cities through gradual improvements in infrastructures, functions, and other dimensions toward the established standards. The third type of eco-city results from post-disaster reconstruction, examples of which include Beichuan and Yushu. These eco-cities focus on meeting national environmental protection, energy conservation, emission reduction, and livability standards from the beginning. Examining Chinese eco-cities reveals six categories: landscape and recreation-oriented cities, green industry-oriented cities, resource-conserving cities, environmentally friendly cities, circular economy-oriented cities, and green consumption-oriented cities. Landscape and recreation-oriented cities operate at a high eco-civilization level. Urban landscapes serve as the primary recreation spaces in these cities, the majority of recreation activities are done outdoors, and ecosystem services are found at the core of urban development. The city of Suzhou, known for cultural landscapes, provides a typical example. Green industry-oriented cities promote green concepts, utilize advanced technology, and establish green industry systems to achieve sustainable social, economic, and environmental development, balance the natural ecosystem, and allow for harmony between cultural landscapes and ecosystems. The city of Kunming focuses on building green industry systems, providing a typical example of this type of city. Resource-conserving cities take legal, economic, administrative, and other measures in production, distribution, and consumption to realize efficient resource use, optimized economic and social benefits with the lowest possible resource consumption, and ensured sustainable socioeconomic development. The city of Beijing has long worked toward energy conservation and emission reduction, providing a representative example of this type of city. Environmentally friendly cities build in consideration of the environment's carrying capacity, obey the laws of nature, deploy green technologies throughout all sectors, and advocate for a green culture and an eco-civilization. The city of Guilin boasts easy access to unique natural landscapes, providing a good example of this type of city. Circular economy-oriented cities are known for good spatial arrangements, sound infrastructure, a clean and beautiful environment, high levels of safety and comfort in life, efficient utilization of materials and energy, and sustainable resource use, as factors reliant upon a circular economy. The city of Guiyang focuses on developing circular economy and so represents this type of city. Green consumption-oriented cities promote green product consumption to further environmental protection and conservation. The city of Dalian has long been promoting green lifestyles and serves as a typical example of this type of city.

Shared Ideas for Eco-City Construction Chinese organizations accumulated some experience in building eco-cities during the eco-city construction wave, leading to some agreements on strategies and points deserving attention. The first agreement is that a clear eco-city development plan is needed as the cornerstone for the eco-city's construction. Objectives and indicators are defined in the eco-city development plan for each phase of each city's development, providing a series of projects that ensure the scientific implementation of the plan. The second agreement dictates that ecological function zones shall play a significant role in eco-city construction. Eco-city development in China distinguishes itself by emphasizing ecological function zoning. The relevant government authority drafts an ecological function zoning scheme based on comprehensive assessment on the current ecological and environmental conditions, ecological vulnerability, and ecosystem services, keeping in mind current spatial arrangements and local ecosystem characteristics. Relevant organizations further study each ecological function zone to determine the ecosystem's characteristics, functions, and trends and resources in need of protection. The studies guide local natural resources extraction, utilization, and protection, providing a scientific analysis for industrial arrangements across the city. The third agreement stipulates that developing a circular economy is critical for eco-city construction. Only a circular economy can minimize the natural resources needed for economic activities and their environmental impact. Chinese society has gradually come to accept the need to achieve sustainable economic growth with the lowest possible resource consumption and environmental costs. A comprehensive circular economy and green industrial development systems are planned into industry policy, industrial arrangements, adjustments to the industry mix, and new industrial parks. The fourth agreement demands that eco-cities are built from a systemic and comprehensive perspective. The former SEPA proposed that eco-city development focuses on key issues, in particular on construction of the four central urban systems, at a site meeting about nationwide eco-district and eco-city development held in Suzhou in 2004. The first central urban system is an efficient production system with low energy and resource consumption and low pollution levels developed by promoting green industries. The second central urban system is a stable, harmonious, and high-quality environmental system achieved through environmental improvements. The third central urban system is a beautiful, comfortable, and harmonious human settlement system comprised of eco-friendly structures. The fourth central urban system is a modern and distinctive green culture system achieved by promoting green culture.

2.1.2 Recent Eco-City Developments in China

Diversification Every city enjoys unique geographic, historical, cultural, and infrastructure characteristics. Local governments should research these characteristics before drafting an eco-city development program so they can account for local conditions and highlight local advantages and characteristics in relevant fields. Such diligence will allow eco-city development to profit from a city's advantages and create a remarkable image for the city. Eco-cities are rapidly built across China. A variety of patterns have emerged in a diversified coexistence that includes garden cities, mountain and river cities, mountain cities, livable cities, forest cities, green cities, and low-carbon cities. These patterns share a common goal – achieving harmony between the built and natural environments and creating an environment supportive of humanity and its growth. Despite the common goal, these patterns do not share the same priorities. These patterns align with the complex and diverse natural and cultural circumstances found across China. Chinese eco-cities can be divided into three categories based on implementation and organizational structure, including international cooperation, ministry–province (or city) partnership, and independent. The SSTEC and the Caofeidian International Eco-city are examples of type 1 international cooperation between China and Sweden, and Chongqing Yuelai Eco-city is the result of cooperation between China and the United States. The Wuxi Eco-city and Guangming New Area in Shenzhen are examples of type 2 partnerships. The Anji District in Zhejiang Province, Huainan City in the Anhui Municipality, and Yanqing and Miyun Districts in Beijing all provide as examples of type 3 independent development. Chinese eco-cities are also grouped into two categories determined by the scope of implementation. Implementation focused on a specific issue, like the focus on concentrated gas and the ecological rehabilitation of coal mining subsidence areas in Huainan, is one possible scope. The other scope focuses on integrated implementation, found in Anji District and many other eco-cities in China. A large number of Chinese cities are pursuing unique eco-city development patterns ideal for their local situation by implementing stylized projects. The Caofeidian International Eco-city is focused on establishing a circular economy in accordance with its strengths; the Wanzhuang Eco-city, part of Langfang City in Hebei Province, is concentrated on creating a green transportation system; the Dongtan Eco-city in Shanghai City is developing eco-friendly farms and wetland parks; Yuxi City in Yunnan Province is developing aquatic ecosystems adjusted to the plateau environment; and the Lanzhou New Area in Gansu Province is establishing a demonstration zone for the rehabilitation of mountain ecosystems and a sports-themed ecological park.

Focus on Low-Carbon Emissions All countries have agreed that alleviating global warming and reducing carbon dioxide emissions are necessary as the world enters the twenty-first century. The 1997 Kyoto Protocol followed by the 2007 Bali Road Map and the 2009 Copenhagen Climate Change Conference chart the approaches countries have adopted to fulfilling their respective obligations to reduce carbon dioxide emissions and achieve set targets. All humans are now searching for low-carbon development patterns. China's agenda must prioritize energy conservation, emission reduction, and environmental protection to address its position as the world's largest carbon dioxide emitter. The term "low carbon" has become the China's eco-city development theme. Over 200 Chinese cities at the prefecture level or higher have announced their intention to become a low-carbon city, including Baoding, Shanghai, Beijing, Hangzhou, Wuxi, Wuhu, Dezhou, Zhuhai, Shenzhen, Xiamen, Chengdu, Chongqing, Wuhan, Nanchang, Guiyang, Jilin, and Harbin, following a general rising trend in China. Shenzhen has announced that it will create a long-term low-carbon development program, improve its carbon trading market, promote low-carbon technology development and application, and cooperate on low-carbon development internationally. Guangzhou unveiled its plans to advocate low-carbon lifestyles, build low-carbon communities and transportation systems, and encourage public organizations to conserve energy. Xiamen has announced its plans to become China's leading low-carbon city by aggressive promotion of an eco-civilization. Baoding has announced that it will lower carbon emissions in a growing number of fields and become a distinctive model for low-carbon cities. Shanghai is working to develop renewable energy and improving traditional energy efficiency. It has identified Chongming Ecological Island, Lingang New City, and Hongqiao Central Business District as low-carbon demonstration zones.

The NDRC launched a pilot low-carbon development program in five provinces and eight municipalities in 2010, including in Guangdong, Liaoning, Hubei, Shanxi, and Yunnan provinces and in Tianjin, Chongqing, Shenzhen, Xiamen, Hangzhou, Nanchang, Guiyang, and Baoding. The NDRC requires these municipalities and provinces to include their respective pilot low-carbon development schemes in their local 12th Five-Year Plan for economic and social development. The Chinese government has defined an integrated, three-stage strategy for building low-carbon cities. Stage 1 (2010–2020) will focus on a "no regrets emission reduction" strategy. Stage 2 (2020–2030) will implement the "alternatives to fossil energies" strategy. Stage 3 (2030–2050) will cover the implementation of the general emission reduction strategy.[2] The "no regrets emission reduction" strategy requires Chinese cities to adopt low-carbon measures including developing new energy industries such as clean coal and new energy vehicles (NEVs) to achieve green industry-based emission reductions; developing various low-carbon or zero-carbon products such as LED lights to reduce emissions through reduced energy use; developing public transportation, improving rail infrastructure, limiting private vehicle use, implementing the fuel-to-gas project, and encouraging walking and cycling to reduce transportation-related emissions; developing green building designs and energy conservation technology for buildings to reduce emissions from the building industry; establishing low-carbon consumption patterns to reduce consumer emissions; promoting energy-efficient technological advances to reduce emissions through new technological capabilities.

Promoting "Microcirculation" and Related Technologies A national conference on urban development and planning was held in the city of Yangzhou in Jiangsu Province in June 2011. Qiu Baoxing, the Deputy Minister of the Ministry of Housing and Urban–Rural Development (MOHURD), use the conference to proclaim that China had entered the intermediate and advanced stages of urbanization, requiring city managers to accept independent organization and replace their early aggressive and forceful demolition and rebuilding with urban microcirculation system restoration. The microcirculation system encapsulates degradation, energy, impact, revival, transportation, landscaping, control, etc., all circulation on the microlevel. The restored system must become the new standard for urban transformation and for planning and construction of eco-cities and urban residential buildings.[3] Micro-degradation solves degrading urban waste concerns at the source to alleviate the impact of urban garbage on development. Micro-energy combines energy supply and consumption with a focus on energy acquisition, utilization, and recycling at the same location. Micro-impact aims to minimize the impact of urban development on local ecosystems, steering urban planning to reduce disruption of surface water and groundwater systems. Micro-revival promotes organic urban revival and

[2] Fan et al. [2].

[3] China Society for Urban Studies [3].

development of urban areas that maintain integrity by preserving history, culture, and memory and ensuring sustainable development. Micro-transportation replaces daily trips that span long distances with more convenient and comfortable micro-trips. Micro-landscaping designs landscapes in parks, on rooftops and other structures, and along streets, moving them nearer to city dwellers, alleviating the UHI effect, and improving air quality. Micro-control refers to digital ecosystem management in every community to scientifically control systems and manage the city intelligently. Many Chinese cities are currently testing and applying microcirculation systems as part of eco-city development efforts. As Beijing's separate sources, the eco-friendly drainage system is an example of a micro-degradation effort; the Taihu New City in Wuxi City has an energy plan that introduces micro-energy; municipal roads in the Guangming New Area and the Niushan Technology Park in Shenzhen City implement micro-impact principles; the Sanli River course improvement program in Qian'an City in Hebei Province boosts micro-revival; Hangzhou's bicycle rental system and Hong Kong's pedestrian system are examples of micro-transport; the Shenzhen Greenway and the Green Balcony program in Beijing are micro-landscaping projects; and the SSTEC Green Intelligence city management system and the Intelligent Community program in the Tianjin Binhai New Area offer micro-control.

Development of Theories and Practices Relevant to Green Building Green buildings minimize resource consumption (including energy, land, water, and materials) and pollution to provide healthy, suitable, and efficient spaces for people and to create greater harmony between people's lifestyles and nature. An eco-city is an integrated system that includes green building technology, and green buildings are an essential component of an eco-city. Developing green and energy-efficient buildings is not only the most important and direct approach to building an eco city but further addresses global energy concerns, environmental crises, and climate change. The attendees at the Eco-city and Green Building Summit held in Chengdu on October 17, 2011 discussed the latest developments in renewable energy applications for buildings and future prospects. The attendees at the Eighth International Conference on Green and Energy-Efficient Buildings & New Technologies and Products Expo held in Beijing on March 29–31, 2012 discussed promoting green buildings to create a low-carbon, livable environment. The conference also presented a variety of the latest technological advances in the fields of green building materials, construction, interior design, renewable energy applications, green lighting, public building energy consumption regulation, wall materials, HVAC (heating, ventilation, and air conditioning,) and intelligent appliances. Both conferences contributed to developing theories and practices for green buildings in China and thus encouraged eco-city development by means of energy conservation in buildings.

Green buildings are an integrated, optimal configuration for comfort and resource conservation. The Chinese government called for increases in the energy efficiency of existing buildings and construction of new, energy-efficient buildings in the 12th Five-Year Plan. Most green building projects are currently focused on increasing energy efficiency through technological and management improvements. China's green building are evolving from nonselective technologies to selective technologies, according to a study by Tian Huifeng et al. Shanghai, Guangdong, and Jiangsu are the leaders in the number of applications for green building certificates compared with other Chinese municipalities and provinces. Water- and energy-efficient technologies are the most frequently applied in green building construction in China, whereas technologies concerning material conservation and operation and management improvements are the least used. Chinese organizations are aggressively promoting the use of solar energy technologies. Geothermal heat pumps, Energy Performance Contracting (EPC), enveloping thermal insulation, energy-efficient windows, energy consumption simulation and optimization, water-conserving equipment, and soundproof designs are the most frequently energy conservation technologies used in green buildings.[4]

Improved Eco-Industrial Park (EIP) Construction and Management The *Report on Low-Carbon Eco-cities in China 2012* mentions that EIPs will help reduce pollution, enhance infrastructure designed for pollution prevention and control, and continually improve the environmental risk prevention and control mechanisms. EIPs will also provide critical support for energy conservation, emission reduction, and environmental safety at the regional and industrial levels and for eco-city construction in China. The MEP, the Ministry of Commerce (MOFCOM,) and the MOST co-issued the *Guiding Opinion on Enhancing the Building of State-Level Eco-industrial Demonstration Zones* on December 5, 2011 in accordance with the *Opinion of the State Council on Enhancing Major Environmental Protection Projects*, the *Notice on Building State-Level Eco-industrial Demonstration Zones*, and the *Administrative Measures for State-Level Eco-industrial Demonstration Zones (Trial)* in order to further implementation of the Scientific Outlook on Development and encourage eco-civilization development. The document urges organization to enhance the sustainability of relevant eco-industrial demonstration zones through conceptual, technical, and managerial innovations; follow the three Rs of reduce, reuse, recover, treat wastes into harmless materials, and encourage the development of low-carbon industry clusters functioning on a circular economy in the demonstration zones; lead sustainable socioeconomic development by demonstrating successful solutions and improving the industry mix in demonstration zones; combine policy incentives and market forces with public engagement to establish long-term development of state-level eco-industrial demonstration zones.[5]

The Chinese government has announced a series of policies for EIPs such as prioritizing approval and initiation of intra-EIP projects designed for pollution prevention and control infrastructure, comprehensive utilization of energy and resources, and a complete eco-friendly industry chain; providing EIPs with financial support and tax incentives; implementing pricing policies that facilitate more

[4]Tian et al. [4].

[5]The Ministries of Environmental Protection, Commerce, and Science and Technology, the *Guiding Opinion on Enhancing the Building of State-level Eco-industrial Demonstration Zones*, December 5, 2011.

efficient energy and resource use in state-level EIPs, such as differentiated electricity and water prices; and developing and improving state-level EIPs' incentives.

Higher Public Awareness Surrounding Environmental Protection Public awareness of environmental protection efforts includes the extent to which the public participate in environmental protection on their own initiative and the public's sense of responsibility for building an eco-civilization. The awareness provides fundamental support that humanity needs in order to eliminate the environmental crisis, prevent and control environmental pollution, and conserve resources. Environmental awareness can be increased through government encouragement, environmental incidents, mass media guidance, and environmental NGOs promotion. Numerous events in 2011 and 2012 suggest that public awareness surrounding environmental protection was increasing in China. The four factors for increasing public awareness about environmental protection can be analyzed as pertaining to the situation in China. Development toward an eco-civilization was highlighted in a report to the 18th CPC National Congress. This report defined a five-in-one national development strategy that pursued development toward an eco-civilization in combination with economic, political, cultural, and social development. The eco-civilization has become an important part of the national ideology, to the point where the public have learned to recognize and accept it. The Scientific Outlook on Development has become the primary guideline of the CPC promoting ideas such as human-oriented action, comprehensive coordination, and sustainability. As these terms have become more common in media reports, environmental friendliness and resource conservation have grown in popularity with the public. Public protests against a Paraxylene (PX) project in the city of Ningbo in Zhejiang Province in 2012 demonstrated a public concern about the safety of the living environment and led to national discussions on environmental impact assessments (EIA). A growing number of media reports on environmental incidents have drawn the public's attention. The *Southern Weekly* has an environmental section focused on going green, environmental protection, and new energies. Sohu, one of the leading Web portals in China, launched the Low-Carbon Stars program, which received support from approximately 100 important players across various sectors and spread many green concepts. Massive media coverage directed the attention of city managers to green GDP, explained technical terms such as PM 2.5 to a considerable part of the public, and introduced low-carbon efforts as a common environmental topic for many urban residents. A variety of environmental events launched by environmental NGOs have contributed to higher environmental awareness and to promotion of an eco-civilization. Environmental NGOs provide relevant information and propose solutions through their events and projects. They have been playing an important role in increasing public engagement in environmental protection, assisting governments in rehabilitating the environment, and promoting environmental legislation. Environmental NGOs have recently begun a new phase, expanding their breadth and reach through the Internet.

2.1.3 Noteworthy Eco-City Development Issues in China

Much progress remains on the road toward eco-city development in China. Many eco-city development incidents and problems emerged across China in 2011 and 2012. These incidents and problems deserve close attention and reflection to provide new ideas and solutions for eco-city development.

A Rapidly Growing Automotive Industry The automotive industry is a pillar of any industrialized country, fueling national economic growth and particular growth in associated industries such as iron, steel, and petrochemicals. The automotive industry has greatly enhanced human mobility, expanding possibilities at work and in daily life. The rapidly growing automotive industry has also produced many negative effects on cities, such as air pollution from exhaust, environmental pollution from oils and coolants, disturbing noise pollution, a growing amount of urban space redistributed for parking lots, travel difficulties due to traffic congestion and the associated time that is wasted, and the global energy crisis triggered by the massive consumption of crude oil. In his book *EcoCities: Rebuilding Cities in Balance with Nature*, Richard Register attacked automobiles, highlighting their negative effects on urban development. He believes that cars are the dinosaurs of our time and are destroying the reasonable and happy structures of cities, towns, and villages.[6] He further asserts that the entire automotive transport infrastructure spanning automobiles, urban sprawl, freeways, and oil produces horrible effects that have taken a dominant role in life on Earth and are the cause of resources exhaustion, habitat destruction, climate change, and species extinction.[7] Register thinks that cities should be compact and designed for humans, not for automobiles. Eco-cities change to city dwellers' production, and consumption patterns can help them end their automobile dependency, eliminating the consumption pattern rooted in fossil energy.

An overview of the global history of urban development reveals a consistent thread along urban development stories – the city's focus shifts from pedestrians to automobiles. Urbanization and automobile saturation are both experiencing immense growth in China today. The number of automobiles is growing fast in every major Chinese city as car prices drop and residents' income rises. China has become a major automobile producer and consumer. Production and sales of automobiles grew steadily in 2011, exceeding 18.4 million units and setting another world record, according to a report on the status of China's automotive industry published by the Ministry of Industry and Information Technology (MIIT) on January 20, 2012. By the end of June 2012, there were more than 114 million vehicles across China, including 86,130,000 private cars, according to a xinhuanet.com report from November 10, 2012. Seventeen major Chinese cities are home to more than one million cars, and Beijing, Chengdu, Tianjin, Shenzhen, and Shanghai each

[6] Richard [5].

[7] Richard [6].

have more than two million cars in their streets. The nonrenewable economic and consumption pattern prevalent in many developed countries that relies on petrochemical fuels and centers on automobiles is not suitable for the huge population base and high population density in China. Before cars become the primary urban transportation mode, aggressive regulatory action should seek opportunities to properly handle urban problems caused by automobiles. Cities should prioritize proximity, and city sizes should remain controlled and compact. Green urban transport systems should give walking a first priority, followed by cycling, railroad, public rail systems, cars, and trucks. Chinese cities involved in pilot energy-saving vehicle and NEV demonstration programs should gather their experience so that more efficient vehicles and intelligent transportation systems can be aggressively pursued.

Infrastructure Development An unusually large flood swept Beijing on July 21, 2012 as a result of the heaviest rain in 61 years. Traffic was interrupted all over the city, many vehicles were submerged, and 77 people were killed as a result of drowning, electric shock, collapsed houses, debris flows, traumatic shock, falling objects, and lightning. Floods were recorded in 62 % of Chinese cities between 2008 and 2010, among which 137 cities experienced at least three serious floods, according to an MOC survey of 351 cities in 2010.[8] These events have exposed urban development problems in China. Investment in the infrastructure networks, especially drainage systems, is far below a satisfactory standard considering rapid urbanization. Rapid expansion of the city has severely damaged and disrupted urban vegetation and natural hydrological conditions. Many cities have not yet developed comprehensive coordinated design plans for rain and flood control with regard to landscaping, water resource utilization, road systems, and administration. A large number of cities lack early warning mechanisms, technologies, and techniques that would be useful in the event of extreme rainfall. The floods have forced new perspectives on how infrastructure relates to urban development. China is increasingly struggling with outdated urban flood control and drainage systems that do not keep pace with rapid urban expansion. Urban expansion has irrationally prioritized what is on the ground over what is underground, leaving no rooms for considerations of urban flood control and drainage infrastructure. City development trends often lead to short-sighted and arbitrary occupation of water areas near rivers and lakes. Cities regularly expand beyond the flood control boundaries or into low-lying areas without employing necessary protection measures. The expansion leaves floods with huge potential to cause great damage.

Urban human settlements include natural, human, social, residential, and support systems, with infrastructure including public utilities such as tap water, energy, wastewater treatment, transportation, communications and IT systems and a physical environment plan composing most of support systems.[9] A scientific, convenient, safe, and reasonable infrastructure support system that complies with ecological

[8] Che [7].

[9] Wu [8].

laws is critical for eco-city development. New urban areas are being built in many Chinese cities, and old areas are reconstructed. Urban planning and road construction must correspond with a network of underground pipes during this construction. This practice will satisfy drainage, sewage, water, gas, heat, electricity, communications, and other needs as the city develops. A systematic plan is needed for urban sewer and other pipe network construction. Careful and precise urban water resources supply, drainage, and wastewater treatment and recycling system designs are also crucial. Advanced rainwater treatment methods and technologies from other countries should be introduced. Damage to infrastructure construction and to the city's ecological footprint and natural environment should be avoided.

Air Quality Monitoring Standard Smog appeared in many cities such as Beijing, Shanghai, Nanjing, and Jinan during 2011 and 2012 pushing discussions of air quality monitoring standards to the forefront. Observed air quality data and public sentiment vary greatly. Pollutant emissions in China are high, regional air pollution problems are severe, and the overall atmospheric environment is grim. Ambient air quality monitoring points in Chinese cities reflect the overall environmental quality. However, not enough monitoring points are available to properly represent the area covered. Monitored items, monitoring networks, the number of point monitored, and the points' locations are unsatisfactory based on human health impact assessment requirements. Embassies in Beijing and some civil society actors in China have requested that PM 2.5 be incorporated in the air quality monitoring standard over the last few years. The State Council approved a newly revised ambient air quality standard at an executive meeting on February 29, 2012 in order to bring ambient air quality assessments in line with reality and citizens' personal experience and to strengthen air pollution prevention and control. The new standard incorporates PM 2.5 monitoring indicators and 8 h ozone concentration limits, thereby officially incorporating PM 2.5 into China's air quality standard. The State Council meeting dictated that the fine particulate matter, ozone, and other substances will be monitored in the Beijing–Tianjin–Hebei region, the Yangtze and Pearl River Deltas, in municipalities directly under the central government's responsibility, in provincial capital cities, and in other key regions in 2012 and that the monitoring effort be extended to 113 key cities in 2013 that serve as environmental protection role models, before all prefecture-level cities or higher engage in monitoring in 2015.

PM 2.5 is a measure of particles which are no larger than 2.5 μm in diameter. PM 2.5 is a primary contributor to smog and carries substances that seriously harm human health because they are small enough to enter directly into the lungs and even into the blood stream through the nasal cavity. The Beijing–Tianjin–Hebei region has suffered frequent smog incidents as a result of growing PM 2.5 pollution in recent years as the number of motor vehicles and coal-fired energy consumption has increased dramatically in the region. Now that PM 2.5 has been incorporated into China's air quality monitoring standard, each big city monitors air quality, indicating the country's determination to strengthen comprehensive air pollution prevention and control and promote continuous air quality improvement by implementing higher standards and more effective measures. The following points demonstrate the importance of air quality monitoring standards to eco-cities. First, obsolete technology and equipment in industries such as electric power, iron, steel, building materials, nonferrous metals, petrochemicals, and other chemicals should be phased out at faster rates in every city. Promoting the clean energy use, relocating heavily polluting enterprises, and encouraging energy-efficient, environmentally friendly technologies in those enterprises will further optimize the local industry mix. Second, raising environmental requirements for market and industry access, implementing more stringent air pollutant emission limits in key areas, and prohibiting construction or expansion of coal-fired power plants except for cogeneration facilities, steel mills, and cement factories are also crucial in improving air quality. Tight control over new projects and closely monitoring environmental law enforcement is pivotal to these ends. Third, motor vehicle pollution control, improved fuel quality, and emission standards form another important tool. Fourth, intercity collaboration on air pollution prevention and control, especially between cities in the Beijing–Tianjin–Hebei region, Yangtze and Pearl River Deltas, and other key areas needs strengthening for effective action. Air pollution warning systems for extreme weather conditions are also advised.

Environmental Impact Assessment (EIA) An EIA analyzes, predicts, and assesses the possible environmental impact of implementing a plan or project so that countermeasures preventing or mitigating adverse environmental effects can be provided and the progress tracked and monitored. An EIA analyzes the possible effects of a project on the environment after it is put into operation and proposes strategies and measures to prevent pollution. The EIA system encourages environmental consideration during planning phases, with a hope to ultimately render human activities more compatible with the environment. The State has strengthened project EIA regulations in recent years and imposed harsh measures against projects that do not comply with environmental law. The State has explicitly prohibited the following: projects that do not comply with industrial policies; projects that will cause serious environmental pollution, diminished product quality, increased energy and material consumption, and projects that do not comply with pollution emissions regulations; projects that do not meet environmental function requirements even if they provide positive economic benefits; and projects located in nature reserves, core zones, and buffer zones. The State further introduced a series of strict rules including tight limitations on projects involving drinking water sources, nature reserves, scenic areas, and important ecological function zones; tight control over projects that will lead to high energy and material consumption and high pollutant emissions; banning projects that will negatively impact technological improvement or domestic demand stimulation; and constraints on regional, industrial, and enterprise development based on total pollutant discharges. Regional EIA approval has been suspended on a regular basis by environmental regulators in accordance with the Water Pollution Prevention and Control Law of the People's Republic of China and the Regulations on Environmental Impact Assessment of Plans. Regional EIA approval suspension allows the environmental authority with in jurisdiction to suspend the approval of all enterprise and regional projects for enterprises and regions that have violated environmental regulation, with exception of projects stimulating a circular economy, until the noncompliant projects have been adjusted to regulation standards. This measure has given the EIA enormous power to adjust China's industry mix and shift its economic growth. The Chinese public has become increasingly concerned in recent years over the possible effects of certain projects on their living environment. Some of these projects were launched secretly, and the responsible government departments and organizations failed to grant the public full access to relevant information or participate in project evaluation and in the EIA process, causing great commotion. An example of this issue can be found when hundreds of villagers in Wantang and other villages in Zhenhai District, part of Ningbo City in Zhejiang Province, went to the district government on October 22, 2012 to complain about a PX project requested by a local oil refinery as part of an integrated expansion project, because it was too close to these villages. The municipal government talked with the project investor on October 28, after which it decided to cancel the PX project, suspend preliminary work for the integrated project, and conduct scientific research about the project. This was one of many incidents against PX projects covering many Chinese cities over the past few years. In 2011, Dalian citizens similarly demonstrated against a local PX project. Earlier, in 2007, Xiamen citizens staged a march to oppose a local PX project. PX is a low toxicity, flammable chemical that causes irritation. It causes irritation to the eyes and upper respiratory tract and has an anesthetic effect on the central nervous system, causing acute poisoning when in high doses. The anti-PX incidents demonstrate two things. First, the NIMBY ("Not in My Backyard") attitude is growing among the public. The "NIMBY effect" refers to strong and sometimes highly emotional opposition against certain projects, such as dumping grounds, nuclear power plants, crematoria, etc., led by local residents due to fear of the projects' potential negative effects on physical health, environmental quality, and asset value. Second, the current EIA system in China needs immediate improvements, most importantly granting the public full participation in the EIA process to provide a greater understanding of relevant issues and offer an opportunity to discover, analyze, and solve problems.

The 18th CPC National Congress established that economic development must incorporate political, cultural, and social development in working toward an eco-civilization and that China's comprehensive sustainable development relies on a national development strategy that incorporates all five components. The Ningbo incident will encourage more cities to use the five-in-one strategy for eco-civilization construction, including an emphasis on EIA. The EIA requires strengthening and alignment with governing laws and regulations. Public information disclosure and access to EIA information, including government commitments at all levels, are also critical. Increased public participation in the EIA process, respect for people's rights to be informed, and careful consideration of public opinion will help EIA's emphasis. Social risk assessment mechanisms for important projects with full consideration of their potential environmental damage against local residents can be established and improved, as can resolute implementation of all of the listed EIA policies, particularly regional EIA approval suspension.

2.1.4 Main Issues Facing Eco-City Development in China

Although China has made impressive achievements in eco-city development over recent years, it still faces many problems and challenges in this field when compared with other countries.

Eco-Friendly Development Is Overemphasized in Central Cities While Regional Interaction Is Overlooked, Especially Urban and Rural Interaction A city and its surrounding region are inseparable – the city is the core of the region, and the region forms a foundation for the city. Materials, energy, and information are continuously exchanged within the city and the region. The exchanges will become increasingly frequent, and the interactions will intensify alongside urban development. Intercity and interregional collaboration will provide benefits to eco-city development. Central cities' eco-friendly development is overemphasized in China's current eco-city development phase, ignoring interregional cooperation and development. Government departments and organizations have not considered the complementary effect of the impact of large-scale regional development and city clustering on eco-friendly development. Urban and rural areas tightly connected and have particularly profound internal correlations in the field of eco-friendly development. A city planning eco-friendly development must consider the development of affiliated rural areas. The city must draft an overall plan for the socioeconomic development of urban and rural areas, paying attention to the complementary effect between eco-friendly industry and agriculture in developing into an eco-city and using the circular economy concept to guide its policies. The city can thus develop holistic green industries and realize sustainable development.

A Lack of Legal Support for Eco-City Development and Ineffective Implementation of Eco-Friendly Standards Eco-city construction has been normalized across Chinese society as the sustainable development strategy was implemented and greater knowledge about the eco-civilization and the Scientific Outlook on Development has become available. The public is now aware of the urgent need for eco-cities. Mandatory constraints and regulations for eco-city development are still lacking however, as both governments and enterprises tend to prioritize economic indicators over environmental indicators. This phenomenon is most pronounced in the following aspects. Economic benefits are overemphasized, and the environment is overlooked in urbanization processes. The city plan contains reference designs for green spaces, building height and density, and infrastructure, but the references are not properly adhered to during urbanization. Urban development management is fragmented and lack unification. Some organizations arbitrarily occupy public spaces regardless of the aesthetic effects on the urban environment or the public interest to meet their own interests. Some enterprises fail to meet relevant pollutant discharge and emission standards, and such environmental pollution remains a major concern in the city. These challenges have prevented efforts to effectively curb the trend of ecological deterioration in Chinese cities.

A Lack of Synergy in Eco-City Development and Limited Public Engagement The comprehensive project of eco-city development requires close cooperation and collaboration among various organizations across society. It cannot thrive if broken down among administrative departments where personnel organize piecemeal implementation. The government must formulate policies based on careful consideration of public opinion to recognize the large role of the public in eco-city development. Policies that fully reflect people's needs can be persuasive and influential enough to gradually urge the public's attention to environmental protection and further personal initiative and engagement, thereby contributing to development toward a harmonious society. Eco-city construction requires the efforts of every citizen. Every person is both responsible for environmental damage and a victim of the resulting damage, linking environmental trends directly to every individual. Current eco-city construction involves government departments and enterprises but suffers from a blatant lack of public participation. Eco-city development will remain vapid unless all people are invited to participate. Strengthening public participation, including that of environmental organizations, environmental education, etc., is of topmost priority in furthering eco-city development. This shift the current eco-city development trend where government are work without much citizen input.

Failure to Consider Local Conditions and the Lack of Focus and Uniqueness Eco-city development outside of China often seeks problem-specific projects with outstanding results. Foreign cities tend to focus on addressing one or two major urban development problems, such as traffic congestion, surface hardening, and garbage pollution instead of attempting eco-city projects all across the city. The practice accumulates experience, develops human resources, educates the public, and builds a desirable image while solving one problem at a time. The city of Curitiba, Brazil, focused on urban transportation and on converting garbage into usable resources; some Japanese cities focused on EIP and circular economy development; and some European cities focused on eco-friendly communities. Eco-city development in China lacks this focus and unique solutions. Chinese cities tend to design and build green spaces without considering local conditions. Many of these cities waste their financial and environmental resources on lawn creation. Landscaping should be adapted to local conditions and focused on mixing and diversifying trees together with bushes, grasses, vines, and flowers. Chinese cities generally suffer from unreasonable spatial arrangements. Local terrain features have changed as urbanization has increased. Excessively built landscapes, uncontrolled scenic development, and other undesirable phenomena have been rampant, replacing architecture unique to the area and leading cities to become increasingly similar. Architectural plans and designs fail to reflect local cultural characteristics. Pressure to be eco-friendly and respect and care for natural topography, landscape, and other living things have caused neglect of the city's humanistic spirit and culture. Buildings consistent with the city's nature and the unified eco-city development plan will provide a desirable role for buildings in urban development. Chinese cities should determine personalized focus and development strategies reflective of the diversity of local natural conditions, economic development levels, societal backgrounds, and other basic conditions as they pursue eco-city development.

2.2 Eco-City Development in China: Principles, Ideas, and Practices

Building eco-cities to model an eco-civilization and modern cities is new to China and lacks established patterns and methods. The gap must be bridged through openness to new ideas and innovation.

2.2.1 Principles

A System's Perspective The city differs from any other biotic community because it is an artificial ecosystem based upon human behavior, the environment, the flow of materials, and social institutions. A sustainable city depends on a sustainable supply of resources; the underlying self-adjustment capacity of the ecosystem; the socioeconomic systems' ability to organize and adjust itself, the coordination among ecological functions such as production, consumption, and restoration; macroscopic regulation and control; intersectoral coordination; and public awareness and participation that cover a broad swath of society. Weakening or enhancing any function will affect the other components and the city's sustainable development. Eco-city development must adhere to ecosystem principle to maintain balance. It requires comprehensive and systematic analysis and assessment of nature, of the economy, and of society as a whole.

Recycling Resource, product, and waste reuse and recycling in the urban system are crucial for enhanced ecosystem support to urban development in light of limits on physical resources in a city. Reuse and recycling in an eco-city are also necessary for long-term survival and development. Urban ecological and environmental problems stem primarily from underdeveloped internal recycling mechanisms for materials and products. The city requires large quantities of physical resources to create a relatively small amount of products, most of which are later discarded or emitted in the form of waste. This inefficient resource utilization also induces a large number of environmental problems. Urban ecosystem's resource use habits must be transformed from a single linear "chain" into a composite "net" for this situation to shift. Conducive recycling channels linking resources and waste inside the city and to its surroundings are needed to improve the city's ecological and environmental effects.

Harmony and Symbiosis Harmony is at the core of regulation geared toward eco-cities. Eco-cities strive to mediate relationships between humanity and nature, cities and the countryside, and the built and natural environment. Failure to properly maintain any of these relationships will negatively affect normal urban ecosystem functions. Eco-cities require coordination and organic coupling of the various components. Harmony and order within any individual component are insufficient to guarantee healthy eco-city development. Harmony and symbiosis spread along all system components are necessary for the system to achieve its optimal status. An optimal symbiotic system will conserve resources and energy and provide multiple benefits for itself. Unreasonable resource utilization and weak component coupling will weaken the symbiotic relationship and lead the system's health and vigor to deteriorate.

Sustainable Endogenous Development The urban ecosystem is self-contained, providing for internal regulation, stabilization, and development within a certain ecological threshold range. Ecological thresholds and ecosystem's carrying capacities draw the line where the effects of human society become too much for a particular environment to bear without harmful changes to human survival and development introduced alongside and detrimental effects to the state and structure of the environment. These effects are measured by scale, strength, and rate. An ecosystem's carrying capacity includes the carrying capacities of resources, technology, and pollution. The capacity depends on the structure and functions of the city and is a combination of objective and subjective factors. An objective carrying capacity exists even if it can be altered through human action. Changes to the carrying capacity will cause structural and functional changes in the urban ecosystem, pushing the ecosystem toward normal or reverse succession. The urban ecosystem can evolve in complexity, optimal energy use, and productivity (i.e., normal succession), leading to sustainable development, if the intensity of urban activities remains below the carrying capacity. Urban sustainable development is otherwise impossible.

2.2.2 Ideas

Eco-city development is a process by which a city's construction and management plan are unified with respect to ecological laws (regarding nature, the economy, and humanity.) This process coordinates various ecological relationships – competition, symbiosis, revival, and self-development – among the social, economic, and natural environments of a city; changes the production and consumption patterns, decision-making, and management practices and culture; and increases ecological and environmental awareness by advocating for corresponding values and advancing institutional reforms and technological innovation. Eco-city development in China aspires to eco-friendly and sustainable socioeconomic development. In order to accomplish that goal, China is establishing a healthy and safe environmental support system rooted in ecological safety and an internationally competitive eco-industrial economy, promoting comprehensive landscape protection and the development of ecological landscapes and concentrating socioeconomic development around an eco-friendly culture.

(1) China is striving to develop eco-friendly industries and establish an efficient production system with low consumption and pollution. Cleaner production and industrial restructuring along with an eco-economy and new industrialization approaches using advanced technology, high economic returns, low resource consumption, minor environmental pollution, and efficient utilization of human resources provide first steps toward eco-friendly industries. Businesses, industrial parks, industries, and regions can help the transition by focusing on circular economy development. Recycling industries in industrial parks and in regions that process material flows, energy, and information can boost circulation on multiple layers of the production process. Communities and organizations can participate by normalizing recycling in society. (2) Another effort focuses on improvements to the environment and establishing a stable, harmonious, and high-quality environmental system. Strengthened EIAs and environmental regulations will prevent introduction of major ecosystem damages caused by humans. Avoiding new environmental damages whenever and wherever possible will further stabilize the development process. Increased environmental rehabilitation efforts and new solutions to environmental pollution and damages are necessary. Local governments' role is to create eco-friendly environments and raise air and water quality within their administered areas by balancing their ecosystems. (3) Our surrounding can be improved by creating eco-friendly human settlements and beautiful, comfortable, and harmonious living systems. Construction plans for urban communities and small towns should focus on the people they are meant to serve and combine modern ideas with traditional culture. Plans for human settlements and living systems that consider local economic conditions and the natural environment will yield beautiful, harmonious, fully functional, convenient, and comfortable results. (4) Creating a modern unique eco-culture focused on the environment will form a basis for future work. Disseminating ecological knowledge, popularizing modern civilization concepts and national cultural traditions, improving environmental protection laws and regulations, and respecting environmental ethics will spread the eco-culture. Improving overall human qualities will further solidify the foundation for regional sustainable development.[10]

Along with these national efforts, eco-cities have a number of developmental goals that will strengthen efforts. (1) A sustainable economy is necessary for long-term development. A sustainable economy abandons the old focus on increasing output by increasing input in favor of a distinctive and efficient knowledge economy. This transformation requires bolstered knowledge and technology for economic growth. Further steps include strengthening horizontal interregional cooperation. The result is a competitive sustainable economic system. (2) A sustainable social system is likewise crucial. In order to achieve sustainability, urban population size must be controlled to maintain optimal density and educational systems that will properly educate the populace. A focus on health and desirable behavior in

[10] Zhu [9].

public values is more conducive than the current focus on making money, as is consumer behavior aimed at conservation and going green. Reliable social security and community services are key. This social system will focus on people and provide harmony between humanity and nature. (3) Eco-city resource systems must also be sustainable. Eco-cities rely on science and technology to promote cleaner production and integrated resource utilization to improve their system, as well as to find general ways to increase resource use efficiency. They can further promote growth and accumulation for ecosystem assets to create favorable survival and development condition for current and future generations. (4) An integral part of eco-city development is assuring the sustainability of the environment. Eco-cities can begin by enhancing environmental governance and restoring and protecting the natural environment. Improved environmental quality and protection and development of natural landscapes are also of key importance. Natural landscapes can be used as a framework for a complex structure that shows the city's unique ecological landscapes on multiple levels, functions, and interconnections. (5) Sustainable eco-city urban function is another piece of the solution. Urban infrastructure can be strengthened and enlarged to eliminate bottlenecks and ensure safety and reliability while fully satisfying the socioeconomic development requirements and people's needs. Effective use of infrastructure services will provide smooth transfer of materials, energy, people, and information across the city. (6) Sustainable management systems allow for proper maintenance of all other systems. Sustainable eco-management must replace traditional regional management. Historical and cultural resources can be used and integrated to benefit management operation, and celebrating urban culture, human spirit, and poetic tone of urban culture provides further material. A mature and integrated sustainable management system covering managerial decisions, cultural heritage, cognitive ability, policy systems, public participation, and regional cooperation is needed to ensure comprehensive functionality.

2.2.3 Practices

Based on eco-city development theories and practices around the world and applied to China, the best developed strategy can be summed up with five practices: five in one, two supports, three belts, four parties, and harmonious development.

Five in One China's best developmental approach will simultaneously focus on five goals: expanding villages into towns, building garden cities, introducing cleaner production, promoting green lifestyles, and ensuring an eco-friendly environment. These five goals pave the path toward integrated and eco-friendly urban and rural development.

Expanding Villages into Towns China's 12th Five-Year Plan proposed building "two latitudinal and three longitudinal axes" along which urbanization can develop strategically. This pattern optimizes eastern coastal cities and lends attention to constructing new city groups in the central and western regions with an environment rich in resources and with a solid carrying capacity. This pattern promotes economic growth and expands the market from east to west and from south to north. The two latitudinal and three longitudinal axes follow the continental bridge and the Yangtze River as latitude markers, and the coastlines, Beijing–Guangzhou line, Beijing–Harbin line, and Baotou–Kunming line, as three longitude markers to create a strategic urbanization pattern that uses major urbanized areas as primary supports and other urbanized areas along the axes as important components. The Chinese government should optimize the development of the Bohai Sea, Yangtze River Delta, and Pearl River Delta regions to form three ultra-large city groups. It should further develop several new large regional and city groups in other important areas such as the Harbin–Changchun zone, the region sandwiched between the Yangtze and Huaihe Rivers, the west side of the Taiwan coast, Central China, the middle reaches of the Yangtze River, the Beibu Gulf (commonly known in the West as the Gulf of Tonkin), the Chengdu–Chongqing zone, and the Guanzhong–Tianshui zone. The strategic plan forces eco-cities to develop in parallel with rural urbanization. Urbanization is a dynamic, natural, social, and historic process wherein rural residents migrate into urban areas, growing the secondary and tertiary sectors in these areas, leading to a growing number of towns and cities, and increasing urban population and urban expansion. Villages' transformation into towns is an important part of urbanization. It captures the urbanization of rural lifestyles and production patterns. This transformation will further transform the economic growth pattern and promote industrial restructuring. It grows alongside industrialization and modernization in China and improves rural areas' industry mix. Population, capital, technology, land, and other production factors will be increasingly concentrated as urbanization continues, producing strong economies of scale and clustering effects that contract with the scattered and empty poverty common in rural areas. The following ideas are useful for promoting village urbanization. (1) Resource, environment, land, population density, and ecological carrying capacity require close attention as China urbanizes. Initial planning and urban management adherent to eco-city development standards carried out by the necessary authorities will help avoid mistakes. (2) The spatial arrangement of small, medium, and large cities deserves the Chinese government's attention as it manages urbanization. The 2010 No. 1 central government document hones the government focus onto development of towns and small and medium cities to boost the rural economy using new villages and urbanization. This represents an important strategic shift in China's agricultural policy. Coordinated towns and city development and positive interactions between urbanization and new villages are crucial. Coordinating agricultural modernization with industrialization and urbanization is a priority in this respect. (3) The spatial structure requires optimization. Reducing spaces occupied by industrial facilities, mines, and rural residential areas while expanding services, transportation, urban residential areas, public facilities, and green spaces ensures a balanced trajectory. Urban sprawl, overubiquitous industrial facilities, and scattered economic and industrial development zones prevent this progress and must be combated. (4) Town and city spatial arrangements can also be optimized. Building of compact urban areas and clearly defining the functions and industrial arrangements of cities around the central city in the region will provide great benefit to the urban system. Ensuring economic links and complementary functions between cities will increase the overall competitiveness of the region. (5) Village urbanization requires a more optimal human population distribution. Controlling population size in the main urban areas of every megacity while encouraging surrounding areas and other cities to absorb floating population balances the population distribution. (6) The industry mix also requires optimization. Advanced and efficient industries with high added value can make up an increasingly larger proportion of the industry mix at the behest of governments. High-tech industries, modern services, and advanced manufacturing should adopt a leading role in economic growth. Further steps by governments include encouraging urban agriculture, water-efficient agriculture, and green organic agriculture; aggressively developing energy-efficient, land-conserving, and environmentally friendly advanced manufacturing high-tech industries that grant local enterprises control over intellectual property rights; pushing the marine industry toward advanced technology and greater added value; and accelerating the development of modern services to promptly form an industry mix led by the service economy. (7) Development patterns in for urbanization projects need optimization. In order to fundamentally transform the economic development pattern, increase clean energy consumption in proportion to total energy consumption, and expand the circular economy, governments should encourage broad application of low-carbon technologies to significantly reduce carbon dioxide emissions, develop stringent energy and water consumption standards as well as pollutant discharge and emission standards, ensure that all garbage is rendered harmless and that sewage is dumped in accordance with the set standards, and strengthen regional environmental regulations as a regional mechanism for collaborative pollution prevention and control. (8) Improved urbanization is also dependent on optimized infrastructure distribution. Transportation, energy, water, telecommunication, environmental protection, disaster prevention, and other infrastructure distribution systems can be better integrated in the region with optimized functionality. (9) Eco-friendly urbanization is necessarily dependent on an optimal ecosystem. Ecological restoration and environmental protection require binding targets, development intensity must be controlled, environmental protection requires increased investment, and stronger environmental governance and ecological restoration are needed to ensure an optimal ecosystem. Cleaner river systems, improved water quality, and protection for arable land, water areas, wetlands, woodlands, grasslands, cultural and natural heritage sites, and open green spaces between cities provide further key steps for improving the living environment.

Building Garden Cities Gardens are beautiful and eco-friendly spaces created using natural and social factors, science, and technology to provide people with joy. Garden cities are an important approach to eco-city development. Building garden cities improves the urban ecosystem and beautifies the urban environment by diversifying plant variety, providing complete green space systems, creating beautiful landscapes, and offering a full range of facilities, so that the general public can rest, walk around, or engage in scientific and cultural activities to enhance physical and mental health in the gardens. Gardens further provide a space to protect, breed, and research endangered species. The Chinese government proposed 12 requirements for the national garden city standard, covering everything from urban green space classification and distribution to measurement indicators. This basic garden city national standard is primarily intended to ensure advanced landscaping indicators, reasonable plans and spatial arrangements, continuous urban landscapes, pleasant environmental quality, unique urban style, urban ecosystem improvements, and sustainable development. The standard requires a reasonable arrangement and comprehensive functions of various green spaces to form a holistic system that will produce ecological and environmental benefits. The standard demands that the urban green space system be linked with landscaping for a larger area of land and for large gardens planned by the city. Natural landscapes, woodlands, and agricultural land in suburbs also receive protection for an integrated urban and rural environment. Government departments and organizations responsible for areas in connection with garden city construction have the following tasks before them. Urban landscaping needs strengthening, and urban green spaces require appropriate growth. Green spaces must be large enough to ensure a virtuous urban ecosystem. The total area of green spaces has become an important measure of the modern urban living environment. Green spaces should account for more than 50 % of the total urban land area according to health science experts. Each person requires 30–40 m^2 of green space, and vegetation should cover 30–40 % of the area to ensure a proper oxygen and carbon dioxide balance in the air. The United Nations has proposed eco-city targets of vegetation covering at least 50 % of urban area, green spaces covering an area of at least 60 m^2 per person, and green spaces in the residential areas covering an area of at least 28 m^2 per person. Landscapes in Chinese cities are usually relatively small for various reasons including historical developments that have led to high population density. Chinese cities need appropriate increases to the area of urban green spaces and vegetation coverage to meet the stated targets. Urban landscaping should be planned and designed according to local geographical features by maintaining the original natural and cultural landscapes wherever possible and minimizing disturbances and damage to the local environment; improving urban green space plans and distribution; effectively coordinating the relationship between urban residents and the environment; systematically designing urban landscapes to match the city's climate and the local residents' living environment quality requirements; developing functional zones across the city and determining green space area distribution, plant varieties and population, and types of biotic communities; and building environmental and ecological control zones that adhere to ecological function zone theories where natural ecosystem characteristics and processes are preserved, maintained, or imitated. Urban landscaping should prioritize the ecosystem and become integral in urban development project planning and implementation. All urban spaces can be ecologically reasonable using diverse landscapes. Urban landscaping should not be confined to aesthetic purposes but should be constructed to also perform important functions like cleaning the urban environment. Relevant organizations should increase the biodiversity of urban landscapes and build reasonable plant populations as well as properly identify plant varieties to create green landscapes with local characteristics typical of the city. Foreign species can be introduced slowly and carefully. Three-dimensional greening across the city is an additional useful tool. Substantial three-dimensional greening can cool and decontaminate the city while reducing noise and promoting an urban ecological balance. Water landscapes' bodies of water play an important role in improving the urban environment. Urban bodies of water increase humidity and regulate it over space and time as well as increase and diversify plants and animal and assist green spaces in forming a better vegetation structure and performing more powerful ecological functions. Bodies of water are also the city's main channel to external ecosystems, river systems being among the city's most important landscape corridors. Urban landscaping emphasizes forestation along local rivers. Green spaces are a city's lungs, and rivers, lakes, and other wetlands are its kidneys. Green spaces need a healthy environment to improve environmental quality in the rest of the city. A river system combined with the land and its biological environment is extremely effective at purification for itself and for the microclimate along the river. A river system reduces the average temperature around it and accelerates air flow, which creates a healthier environment for humans. Rivers provide a habitat for a variety of urban species and purify water. An ecologically resilient river system provides a network of green channels for flood storage and drought mitigation. River's functions are closely related to its neighboring ecosystems so that the isolation from other ecosystems (e.g., by hardening a river's bank slopes with concrete) reduces inter-ecosystem exchanges and purification leaving "dead" water. A city gains its beauty, fresh air, and a pleasant environment from forests and bodies of water. Forests combine with water to facilitate the virtuous urban ecosystem cycle. Urban afforestation provides another key component. Urban forests are becoming an urban landscaping norm. "People in the city, city in the forest" sums up a twenty-first century urban development target. A design philosophy centered around human lies at the core of an eco-friendly urban environment – creating living spaces wherein the city is integrated with forests to establish harmony between humanity and nature. Urban afforestation espouses this harmony and uses green spaces to build a network in a green city. A complete urban forest ecosystem will thoroughly improve the urban environment and ensure material and energy supplies. Urban forest landscapes must be agreeable, friendly, culturally educative, and comfortable to attract people. Efforts must also remain compatible with nature. Relevant organizations should ensure spatial arrangements and plans to match the city's topography and its style, structural characteristics, and spatial attributes. Urban forests should incorporate the structural characteristics of natural vegetation, landscapes, and biotic communities. They should be self-sustaining and comprehensively beneficial. China's twenty-first century sustainable forest development strategy prioritizes urban afforestation. China launched the "build garden cities" campaign in 1992. Twelve cities including Hefei, Zhuhai, Hangzhou, Ma'anshan, Zhongshan, Shenzhen, Weihai, Dalian, Nanjing, Xiamen, and Nanning are currently listed as garden cities. A garden city development wave is currently seeping even further across China. Green space networks incorporating cities and suburbs are important in further development. Chinese urban areas typically have dense buildings, high population density, and small green spaces. Chinese suburbs have comparatively small populations and large natural vegetation areas, including forests, meadows, rivers, lakes, marshes, etc. Large tracts of farmland in the suburbs provide immense ecological benefits unavailable in urban areas. These resources should be well protected and utilized. Governments can capitalize on suburban ecology by promoting farmland shelter belts and bridging the gap between urban and suburban green spaces with lines of plants grown along roads. This practice will complete the green spaces network, create green channels between urban and suburban areas, and supply fresh air for the city to mitigate urban air pollution. Rational plans for the urban green space system covering green points, lines, and areas are needed to create garden cities. The established boundaries for the green points, lines, and areas must then be respected. The boundaries must then be incorporated into green spaces. A green space system that matches urban characteristics is needed from relevant government department and organization to create a livable garden city.

Achieving Cleaner Production Cleaner production was defined by the UNEP in 1990 as "the continuous application of an integrated environmental strategy to processes, products, and services to increase efficiency and reduce risks to humans and the environment." China's Agenda 21 defines cleaner production as "practical production methods and measures that meet people's needs, enable rational natural resource and energy use, and protect the environment. It implements an integrated pollution prevention strategy for production processes and the entire lifecycle of products." Cleaner production conserves production energy and raw materials, eliminates hazardous materials, minimizes the use of toxic and hazardous chemicals and raw materials, and eliminates pollution and harms to humans and the environment. Cleaner production further improves processes, corporate management, and procedures, implements closed-loop material and waste recycling, and reduces waste, converting it to new resources or harmless substances. Cleaner production implies that the city effectively changes the socioeconomic development pattern and actively promotes circular economy growth by developing eco-friendly industries. The following steps will begin the process toward cleaner production. (1) Circular economy concepts should be used to build urban EIPs and accelerate industry mix optimization. Specific measures may include developing energy-efficient high-tech industries with a high added value such as electronics, IT, biomedicine, new energies, and new materials and utilizing advanced technologies to transform traditional industries and to enhance resource efficiency and environmental protection; accelerating energy and water conservation developments; developing industries that use renewable resources and absorb industrial wastes and strictly limit the expansion of polluting industries; implementing tight controls over discharge and emission of industrial pollutants, eliminating obsolete processes, technologies, equipment, and products that waste resources and pollute the environment, and gradually molding an industry mix that is conducive to saving resources and protecting the environment; and developing industrial parks in existing urban development zones and industrial areas following cleaner production policies, regulations, and technical standards and encouraging enterprises that will role model high cleaner production standards. (2) Environmentally friendly products and processes that enable energy conservation, cleaner production, and pollution prevention and control require further research and development along with development of the environmental protection industry. Specific measures may include building development, production, and testing centers for technologies and products critical for environmental protection; supporting enterprises and groups at the foundation of the environmental protection industry; nurturing environmental protection industry centers; using specialization and market principles to promote an environmental service system that covers financing, project implementation, facility operation, technology consulting, information services, and personnel training; increasing the share of environmental services in the environmental protection industry; and improving the quality and competitiveness of environment services.

Living Green Going green stresses innovation and changes to human civilization and traditional industrial thinking to discover lifestyle changes and developmental ideas that can solve environmental problems. The green lifestyle is compatible with ethic values centered on sustainable development. The immediate goal is to promote a low-cost green option that will appeal to people rationally and provide them a happy life that is adapted to societal expectations. The following steps pave the way toward a green life. (1) A more prominent green culture is needed in society. Education about resource conservation and environmental protection should be promoted across society by creating green schools, communities, businesses, hospitals, shops, hotels, etc., and by using a variety of channels such as the media, the Internet, and display spaces to increase the public's environmental awareness and sense of environmental responsibility. Forums, seminars, contests, and other activities on occasions such as the World Environment Day, Earth Day, World Water Day, "Protect the Mother River" Day, Technology Week, and Arbor Day are also useful outreach tools. (2) Green living requires a livable community. Livable community standards and residential areas that concentrate on eco-friendly options during location selection, design, construction, residence, and services are key to a livable community. Governments should aggressively promote eco-friendly residential areas that conserve energy and land and further encourage the use of eco-friendly building materials and of water- and energy-conserving appliances as well as garbage separation and recovery. Community service centers, environmental protection facilities, and other infrastructure and similarly important. (3) A further step involves creating and eco-friendly industrial culture. Businesses and industrial parks can develop an eco-friendly culture by embracing emerging industrial systems and locally distinctive industries. Governments should assist individuals and organizations by promoting environmental awareness and environmentally friendly behaviors, conserving resources, and recycling wastes. Governments should also guide enterprises toward green products, people-oriented services, and environmental protection. Enterprises should establish a corporate culture that supports green technology innovation. Governments should encourage enterprises to design eco-cultural identity projects and build a green business image through product lifecycle management (PLM) and cleaner production implementation. (4) A green consumption concept must be popularized. The Scientific Outlook on Development and concepts such as environmental friendliness, resource conservation, and green consumption should become rooted across society by popularizing ecological knowledge and nurturing eco-friendly attitudes and values. Low-carbon lifestyles should shape consumption through green design, production, packaging, marketing, and consumption method to minimize its negative environmental effect.

Building an Eco-Friendly Environment An eco-friendly environment requires environmental protection that focuses on pollution prevention and control along with ecological safety, as well as harmony between humanity and nature in the environmental system. Clean water and air and safe foods are necessary for this positive living and working environment. The following steps are necessary to lay the foundation for an eco-friendly environment. (1) Environmental protection should be developed into a network. Specific measures may include factoring local conditions in decisions; building and protecting a variety of ecological function zones; establishing ecosystem protection zones with complementary natural and built environments; protecting wetland resources and implementing wetland ecosystem restoration projects through legal, technical, economic, administrative, and other means; enhancing ecosystem protection provided by forests by nurturing and protecting forest resources; and protecting biodiversity and ecosystem safety and stability of wildlife habitats.

(2) Pollution should be prevented and controlled in cities. Specific measures may include implementing a robust environmental access system, developing municipal pollutant discharge and emission standards at levels stricter than their national counterparts; strengthening industrial governance in pollution prevention; implementing clean water projects to reduce water pollution, blue sky projects to reduce air pollution, and low noise projects to reduce noise pollution, promoting circular economic growth; establish a garbage classification and sorting system alongside a recycling network system; reducing solid waste pollution; and encouraging resource reuse.

(3) The urban environment should be focused on people's needs. Specific measures may include using historical and cultural resources such as historic buildings and sites; protecting and promoting the historical, cultural, and folkloric characteristics of the city; and improving urban culture. The Chinese government announced in the 12th Five-Year Plan that it would implement an ecological safety strategy establishing "two zones and three belts" to further a prosperous, democratic, harmonious, modern, and socialist China and to ensure China's sustainable development. China will follow the strategy to build an ecological safety system using the zones of the Qinghai–Tibet Plateau and of the Loess Plateau joined with Sichuan and Yunnan as two ecosystem protection zones. Another zone will be framed by the northeastern forest, northern desertification prevention projects, southern hill and mountain belts, and major rivers. Other ecological function zones critical for the country will be used as important supports, and development zones will be largely excluded from the system. Efforts in the Qinghai–Tibet Plateau ecosystem protection zone will focus on protecting diverse and unique ecosystems and ensuring they

properly supply water to major rivers and regulating climate. Efforts in the Loess Plateau, Sichuan, and Yunnan ecosystem protection zone will focus on preventing soil erosion, protecting natural vegetation, and ensuring the zone maintains the ecological safety of the lower and middle reaches of the Yangtze and Yellow Rivers. Efforts in the northeastern forest belt will focus on protecting forest resources and biodiversity and ensuring the maintained ecological safety of the Northeast Plain. Efforts in the northern desertification prevention belt will focus on developing the shelter belt, protecting grasslands, and preventing sand movement. It is currently impossible to significantly improve the desertified land, but the relevant authorities should ban any human activities that may damage the local ecosystems. The Chinese government should work to ensure that shelter belts in North, Northeast, and Northwest China maintain ecological safety. Efforts in the southern hill and mountain belt will focus on restoring vegetation, preventing soil erosion, and ensuring South and Southwest China's ecological safety.

Integrating Urban and Rural Areas in Eco-City Construction Urban development in China modernizes cities, urbanizes the countryside, and integrates urban and rural areas. Eco-city construction must be coordinated with a new countryside and an eco-friendly integration of urban and rural areas. Many eco-cities outside of China incorporate urban areas and their subsidiary rural areas. Many Chinese cities currently overemphasize areas around the city in eco-city development plans but overlook connections between the city and other areas, especially the ecological interactions with rural areas. An eco-city development plan must look beyond the urban area to urban–rural integration in order to address this problem. The eco-friendly integration of urban and rural areas will allow a complete use of complementary effects gained through urban–rural interaction. Urban and rural eco-friendly integration originates from Sir Ebenezer Howard's garden city theory developed in the United Kingdom. Howard presented this theory in his book *Tomorrow: a Peaceful Path to Real Reform* published in 1898. Howard wrote that "town and country must be married, and out of this joyous union will spring a new hope, a new life, a new civilization," to highlight urban–rural integration as the core of his theory. He believed that the town, the country, and the town–country functioned as three magnets attracting people to live there as illustrated in his famous Three Magnets diagram. He suggested the town–country magnet, or the garden city, as the ideal city, which draws urban and rural advantages. "A garden city is a city designed to provide the population of healthy living and working; the dimensions should be just sufficient to allow the full development of social life, surrounded by a rural belt, soil in its entirety being public property administered or held in trust on behalf of the community." Howard envisaged a garden city that includes both urban and rural components. The garden city is an eco-city on several bases. The city and the countryside surpass simple coexistence to form an organic whole. Coordination between urban and rural areas promotes the entity's growth and establishes a healthy mechanism for internal balance. Both recognize a city's developmental limitation. When a city reaches a certain size, it is restricted, and a new growth is shifted to a neighboring city. A permanent rural belt between these two cities aids in this process. Size restrictions favor higher quality for work and life in the city. Howard's garden city covers only 6,000 acres, housing about 32,000 people. The city center covers 1,000 acres and houses 30,000 people and is surrounded by the remaining 50,000 acres of agricultural land, where 2,000 people live. Howard's garden city and eco-cities both hold that the land and the community belong to the people and not to political or religious leaders. Community facilities are arranged for convenient use by residents. Government departments and organizations must advance two goals to achieve eco-friendly urban and rural integration – strengthening eco-friendly human settlements in the city and developing the new areas in the countryside with preserved ecosystems. Eco-friendly human settlements in the city need to be safe, convenient, comfortable, and focused on people's needs, as well as promoting harmony between humans and nature and society. This can be achieved by optimizing urban spatial arrangements and improving urban infrastructure and urban and rural communities. A number of specific initiatives may be implemented for these purposes. (1) River, lake, wetland, and other bodies of water development plans can be bolstered, along with human settlements surrounding rivers and other bodies of water, to form livable and integrated urban and rural eco-communities. (2) Greening plans along with planting trees and plants cultivate livable and integrated urban and rural eco-communities around the green spaces. (3) Resident migration from populous central cities to peripheral villages and towns will better make use of open spaces, fresh air, and the beautiful and peaceful environment found on the periphery, forming livable eco-communities near farmland. Building a new countryside with preserved ecosystems accelerates rural infrastructure construction, protects rural drinking water sources, improves the rural energy mix, and comprehensively improves the rural environment. Farmer's quality of life will benefit from these changes. A number of specific initiatives may be implemented. (1) Rural settlements require rational planning to avoid inefficient land use and relatively poor rural infrastructure and service facilities that result from overly scattered rural settlements. A six-level city–country system, including primary and secondary city centers, new cities, the central town, general towns, the central village, and grassroots villages, will improve urbanization. (2) Urban infrastructure expansion into rural areas including rural road building; rural infrastructures projects such as water, gas, and heat supply and sewage and garbage treatment facilities; rural drinking water safety projects; and information technology applied in agriculture and rural areas will further rural development. (3) Biogas, solar, wind, biomass, and other clean energies and renewable resources can transform the countryside of introduces. Environmental protection in rural areas and the rural energy mix also need improvement. (4) Landscaping in rural residential areas is not optimal. Rural drinking water and latrine quality are too low, and human health is further endangered by animals housed alongside human and improper river courses and ditches in villages. A village resource recycling system that integrates straw conversion, biogas generation, organic fertilizer use, and eco-friendly farming benefits both farmers and the environment. These steps will clean the home environment, use resources more efficiently, and turn waste to harmless agricultural production, leaving the countryside much more eco-friendly. (5) New towns and villages should be beautiful and eco-friendly.

Two Supports An eco-city has two primary supports for production and life in the city that enable cleaner production and green lifestyles – EIPs and green communities.

Developing EIPs to Enable Cleaner Production in the City Governments have advocated and popularized a circular economy in the wake of eco-city development. EIPs are a standard tool used by the international community to implement circular economy concepts. The CPC declared at the Fourth and Fifth Plenary Sessions of the Sixteenth CPC National Congress that China would develop a circular economy to adjust the industry mix and distribution and to change the economic development pattern. China has begun building EIPs using circular economy concepts to bolster the recycling economy. Six batches of 389 state-level EIPs have been built so far. Many now serve as circular economy examples and support eco-city development across China.

EIPs are new industrial parks that address cleaner production, a circular economy, and the principles of industrial ecology. EIPs connect factories and businesses together using logistics or energy flows to form a symbiotic industry combination where resources and by-product are shared and exchanged. Wastes or by-products at one factory become raw materials or energy sources for another factory. Producers, consumers, and decomposers mimic natural systems to form a circular path in this industrial system and create a closed-loop material cycle, use energy on numerous levels, and minimize waste. EIPs provide economic, ecological, and social benefits by integrating environmental protection and sustainable development into industrial development. EIPs are significant for many reasons. (1) EIPs emphasize cooperation and overall interests. An EIP creates a community of manufacturing and services that produce environmental, economic, and social benefits through cooperation among enterprises across the industry chain. (2) EIPs benefit from systematic planning. Systematically integrating material, energy, and information flows can increase an EIP's operational efficiency and allow infrastructure to be shared. (3) The environment is emphasized in industrial production. The entire industrial production process is planned and designed to match natural ecology so that it integrates into the whole natural system organically and increases harmony with the environment. (4) Sustainability is emphasized. EIP development is sustainable because it minimizes resource consumption and maximizes environmental protection through resource recycling and cascade energy flows. Companies in an EIP implement cleaner production in a transition from traditional passive responses to pollution toward preemptive and thorough pollution control, and they replace old and inefficient production methods known to pollute and consume resources and energy heavily with cleaner, efficient methods that do not pollute as much and reduce resource and energy consumption. The systematic construction of an eco-friendly industry chain has fueled the development and application of technology that conserves energy, recycles waste, and is environmentally friendly, providing a strong technical support system for eco-city development. Chinese cities need more new EIPs as part of urban development. Industrial parks with better spatial distribution, consolidated industrial enterprises across the city, and more examples of a circular economy are also positive intermediary steps.

Building Eco-Communities for a Green Life in the City An eco-community is a green and sustainable community. An eco-community emphasizes integrating ecological relationships between human settlements and the environment with an organic combination of households, buildings, infrastructure, the natural environment, and community-based social services. Howard's garden city theory imagines an ideal city with a strong socioeconomic environment and a beautiful natural environment. This theory has led to the eco-community concept and began reflection on relationships between humanity and nature in human settlements. The *Machu Picchu Charter* published in 1977 defines the goal of urban planning as creating a multifunctional living environment that emphasizes harmony between the living and natural environments. China has increasingly sought environmental plans and designs for residential areas in the twenty-first century. The State has introduced many residential area construction policies and guidance documents, such as the *Points of the Comfortable Housing Project*, the *Guidelines on the Planning and Designing of Demo Urban Residential Areas in the Housing Technology Industry Program for Moderately Prosperous Urban and Rural Areas* (2000), and the *Points of and Technical Guidelines on the Construction of* Green *Residential Areas* (2001). These documents suggest an invigorated focus on residential environmental planning in China that is evolving toward eco-community planning. China is progressing in eco-community development with large-scale projects like the Guangzhou Science City, Hangzhou Ecopark, and the transformation of older urban areas in Shanghai and Xi'an. Green space system design, pleasant and eco-friendly spaces throughout the ecosystem, land and resource conservation, general environmental protection, energy conservation, waste treatment rendering it harmless, waste reduction, and waste recycling all work together to form a green regeneration mechanism for an eco-community. Eco-communities function like cells in an eco-city to promote harmony between humanity and nature and increase people's ecological awareness. Eco-communities require people to protect, utilize, and manage natural resources and cannot tolerate ecosystem damage or environmental pollution. Planners, designers, real estate developers, government departments, community residents, and property management service providers (or community resident committees) each have their role in developing eco-communities.

Three Belts The natural, agricultural, and cultural ecosystem belts are each necessary for urban development. Proper plans and spatial arrangements for these three belts will promote the overall development of the urban eco-civilization. They should be built within each city and among city belts, groups, and circles.

Building a Natural Ecosystem Belt to Optimize the Urban Environment A natural ecosystem belt in the center of the city improves the urban environment. There are several major tasks to accomplish this goal. (1) Protecting forests, developing the forest industry, improving the quality and ecological functions of forests, and developing forests' role in green protection will guarantee the forest contribution.

(2) Building and maintaining urban green corridors and protecting wildlife habitats by preserving river corridors, strip-shaped city parks along rivers, and lines of plants along urban roads will provide necessary ecosystem space. (3) Protecting native plants and biodiversity in urban areas by rescuing and protecting rare and endangered wildlife, protecting original habitats, and creating new habitats will improve the ecosystem's resiliency. Plant communities specific to certain zones should be protected during urban development. Native tree species play a leading role in local natural landscapes and reflect the regional natural ecosystems. Native species and the regional stable plant communities require protection, and foreign species should be introduced sparingly. (4) Protect rivers, lakes, and wetlands and developing environmental protection plans and comprehensive management systems for river basins will ensure their protection. (5) Tourism resources also require protection. Natural landscapes that look artificial, tourist spots that look like cities, and degraded or damaged tourist resources will harm this important industry.

Building an Agricultural Ecosystem Belt to Ensure Urban Sustainability China has been implementing an agricultural strategy focused on 7 areas and 23 zones as part of the 12th Five-Year Plan. The seven areas include the Northeast Plain, North China Plain, Yangtze River basin, Fen-Wei Plain, Hetao Irrigation District, South China, and Gansu–Xinjiang region. The strategy also covers farmland and agricultural areas that function as important components. High-quality rice, special corn, soybean, and livestock industry zones will be built in the Northeast Plain; high-quality special wheat, high-quality cotton, special corn, soybean, and livestock industry zones will be built in North China; high-quality rice, high-quality special wheat, high-quality cotton, canola, livestock, and aquatic industry zones will be built in the Yangtze River basin; high-quality special wheat and special corn industry zones will be built in the Fen-Wei Plain; a high-quality special wheat industry zone will be built in the Hetao Irrigation District; high-quality rice, sugarcane, and aquatic industry zones will be built in South China; and high-quality special wheat and high-quality cotton industry zones will be built in the Gansu–Xinjiang region. This national strategy urges eco-city development to include agricultural ecosystem belts around the city. (1) Each city should use their unique ecological resources to develop eco-agriculture by building green and organic food production centers, establishing a green agricultural production and processing certification system, and increasing green agricultural production. (2) Other features include eco-agriculture development based on biogas, a rotation pattern shifting between grains and cash crops, and the use of pesticides, fertilizers, and plastic films to reduce nonpoint source (NPS) pollution.

Building a Cultural Ecosystem Belt to Promote the Development of the Urban Ecosystem The cultural ecosystem belt in an eco-city encompasses people's cultural values, ways of thinking, and living conditions. This belt is a social living environment composed of systems, institutions, mechanisms, and organizations. The urban cultural ecosystem belt is complex, covering three levels. The first level regards physical objects such as urban buildings, public facilities, cultural heritage sites, artwork, and inventions created during urban development. The second level regards cultural development, including intangible cultural heritage, values, ethical conduct, and religious beliefs, as well as the cultural and artistic heritage embedded in the hearts of city dwellers. Humanism, urban culture, and urban ideology reside at this level. The third level regards management and institutions along with rules, regulations, policies and laws, management mechanisms, and social organizations that can restrict and regulate urban residents. These three levels interact with each other and form the bulk of the urban cultural ecosystem belt.

The cultural ecosystem belt is critical to eco-city development. The most important difference between cities and the natural environment is the cultural environment present in cities. Eco-cities cannot advance to long-lasting development that addresses the deeper identity of eco-cities without a coordinated cultural ecosystem. The cultural ecosystem belt that is inherent in urban development is crucial for eco-city development. People's initiative and cultural strength serve as the inner spiritual core and cultural root of urban development due to their predominant role in the process. An organic balance between natural and cultural ecosystems in the eco-city can be realized by developing urban culture, harnessing people's creativity and wisdom, and promoting positive interactions between cultural and natural ecosystems, thereby protecting natural ecosystem resources and boosting eco-city development. The cultural ecosystem provides cultural support for eco-city development and its growth motivates urban development. A city's historical and cultural heritage can spawn a culture unique to that city and furthers urban culture industry growth. Improved values, stronger public ecological awareness, urban cultural buildings, and intangible cultural heritage culturally educate people in subtle ways as part of the cultural ecosystem. Cultural inheritance, development, and innovation improve a city's internal qualities and show its distinctive charm. A strong cultural ecosystem is conducive to protecting the natural environment of the city. The strong spiritual power inherent in the cultural ecosystem will guide the public to take initiative in protecting the environment and unite people in urban development. Developing systems, policies, laws, regulations, institutions, and organizations involved in managing the relationships among social, economic, and natural ecosystems protects natural ecosystems, including an eco-city's intangible environment and creates a people-oriented eco-city with a beautiful, livable environment. What must be done to create a cultural ecosystem that guides and supports eco-city development? Urban history and culture must be fully recognized and incorporated. The urban cultural ecosystem includes unique cultural elements that are accumulated during a city's evolution, such as lifestyle, traditions, and customs, which can be observed around the city at historic sites, nationally unique buildings, and cultural activities with national characteristics. The traditional character of the city is important to maintain, as is protecting historic sites. Maintaining the city's culture should not rely on preserving traditional streets and nationally unique buildings but should also promote distinctive history and culture, hold various cultural activities, and visually present traditional culture and virtues. The cultural charm of an eco-city can also be found in unique local cultural elements. The municipal government should instruct relevant organization to protect traditional culture and use it to enrich emerging modern culture. The process also requires a humanistic core for the city. The city thrives on humanism, which provides spiritual support for urban entrepreneurship. Establishing the city's humanistic core at this stage of eco-city development requires people-oriented actions, harmonious development, a harmonious socialist culture, enriching cultural activities for citizens, improving the cultural character of the city, innovation, cultural literacy, elegance, and inclusivity, which are indispensable to the urban cultural ecosystem. Intangible cultural heritage must remain in use and be protected. Intangible cultural heritage differs greatly from culture by only incorporating the nonphysical elements of the cultural ecosystem. Protecting intangible cultural heritage is essential for the development of the urban cultural ecosystem. Intangible cultural heritage needs protection to promote sustainability, even against economic interests. Laws and regulations can ensure that intangible cultural heritage is developed, used, and protected from local interest groups interested in exploiting it for their own interests. Efforts should be made to restore diversity and vitality to intangible cultural heritage. Protecting intangible cultural heritage should incorporate the government as well as public engagement and awareness around its importance. Including the public will increase coordination between protection and development. Greater ecological awareness among the public is key. The urban cultural ecosystem's sustainability and development toward a socialist eco-civilization depend on the public's ecological awareness. The urban cultural ecosystem requires a public awareness of its roles, responsibilities, and obligations in the ecosystem and demands that people should adapt themselves to their surroundings as needed. A new set of values conducive to harmony among human beings and in their relations with society and nature is also needed. Governments should encourage public participation in environmental protection and development, promote a harmonious eco-civilization, advocate for environmental ethics and healthy lifestyles, and motivate the public to independently protect the environment.[11]

Four Parties Eco-city development plans and strategies require a strong implementation mechanism. China's eco-city development is led by the government, in contrast with various mechanisms that have emerged in the international community, such as international cooperation, government and business partnerships, and community-driven development. International experience dictates that governments, businesses, NGOs, and the public each have an important role in eco-city development. The four parties should work together to form urban development and management mechanisms that effectively coordinate multiple sectors.

Governments Lead Eco-City Development Governments should coordinate and guide eco-civilization development and provide planning and technical guidance. An eco-friendly government is necessarily collaborative. Social systems must work together to promote harmony between humanity and nature. Governments should coordinate among these social systems, strengthen central and local administration and integrated cooperation, and implement holistic policies to ensure system-wide interests are met. Consensus on environmental protection and PES focused on ecological interests is needed. Governments should coordinate environmental,

[11] Ye et al. [10].

economic, and social development, prioritizing the environment. They should further a balanced view of developmental activities to balance economic, social, and ecological relationships related to human development. The central government should plan for nationwide eco-civilization development, summarize experiences and lessons, and rework plans to best meet eco-civilization goals. It should consider the country's ecological reality and needs in the implementation scheme to guide nationwide eco-civilization activities and ensure efficient and effective coordination among all levels and sectors. Governments should provide policy and institutional support for eco-civilization development. Eco-friendly institutions that adhere to the legal system are critical to ensure the compliance and orderliness of eco-civilization developmental activities because the institutions supplement the legal system. Many factors affect eco-civilization development, and a lack of a public administration mechanism that favors eco-civilization development is an important factor in China. Economic development and environmental protection are unbalanced in some regions of China where local governments overemphasize GDP growth and ignore environmental and resource protection. Some local governments evaluate performance on economic growth and ignore holistic development. This phenomenon has adversely affected eco-civilization development, which requires institutional improvements and innovation to eliminate long-term institutional barriers blocking environmental protection and to harbor institutions conducive to environmental protection. Clearly define developmental ideas and objectives that adhere to essential requirements for eco-civilization development are needed. Governments should establish and improve institutions to favor eco-civilization development and provide support policy and institutional support. Governments must guide development toward eco-civilization goals. Governments should guide the public to systematic participation in environmental protection and encourage various public environmental movements and activities to raise environmental awareness and motivate independent environmental protection. Governments should encourage consumption and lifestyles changes, the resource and energy conservation, the natural environment protection, thrift, and waste recycling. Enterprises and other organizations also need governmental guidance toward environmental information transparency. Access to EIAs can be improved by ensuring that project managers disclose information and promptly provide it to the public to protect the public's right to know and be involved. Advisory bodies and public information networks can provide the public access to timely information on eco-civilization development that will permit people to monitor decision-making processes. Governments should ensure environmental decisions are democratic and strictly adhere to laws and regulations. They should also provide guidance for green consumption transformation, eco-friendly industry growth, and eco-friendly marketing, logistics, and ecological education systems.

Companies Fuel Eco-City Development An eco-civilization requires companies with defined ecological responsibilities that guide operations and business management to achieve sustainable development involving humans, companies, ecosystems, and society. Companies can promote eco-city development by establishing a corporate ecological responsibility system that includes ecologically responsible production and products; ecological responsibility toward nature that abandons traditional values that only emphasize humans' rights to nature in favor of value that includes humans' obligations to protect the natural environment and that ceases robbing nature of its resources during exploitation in order to protect it; ecological responsibility across the market with green market approaches, green products, compliance with environmental protection rules and regulations, and healthy products that meet market needs; and ecological responsibility to the public through shared ecosystem resources and environmental benefits and through proper responses to ecological consequences. It is unjust for the current generation to satisfy their interests at the expense of developmental opportunities for future generations. A corporate environmental management system will support an eco-friendly corporate culture. Environmental guidelines that outline corporate environmental understanding and goals are needed. They include corporate commitments to pollution prevention; observing and improving environmental laws, regulations, and requirements; and creating a framework for setting and reviewing environmental objectives and indicators. Companies should get ISO14001 certification. Certified companies will benefit by establishing a good corporate image and receiving recognition across society; eliminating green barriers to international trade and increasing access to international markets; conserving raw materials and energy while reducing waste to lower costs; and boosting competitiveness through green products. Companies should also establish green corporate values. They need strong green corporate systems embedded in their culture and increased ecological awareness and goodwill among executives. Companies should improve corporate values to contribute to sustainable, environmental, economic, and social development. They should raise the employees' ecological awareness through trainings, seminars, ecological improvement projects, active participation in environmental protection, and public and cultural activities conducive to sustainable development. Companies need corporate ecological management and evaluation systems. A green code of conduct is also needed. Companies can disseminate information on corporate environmental awareness to educate employees and establish a green code of conduct. Corporate ecological and environmental ethics help mediate between corporate interests, public interests, all living things in nature, and future generations. Companies should adopt eco-civilization values at the core of the corporate culture and abide by corporate environmental ethics that respect public interests, natural life, and future generations. These steps lead to a first-rate company. A green corporate image can be crafted through green production, marketing, and management. Chinese enterprises should work together to help China's industrial sector go green, promote efficient corporate resource use, and vigorously promote cleaner production to better conserve resources.

NGOs Contribute to Eco-City Development Many diverse Chinese NGOs are active in environmental protection. The most famous include the Friends of Nature (FON), Global Village of Beijing (GVB), Green Earth Volunteers (GEV), China Small Animal Protection Association (CSAPA), China Environmental Protection

Foundation (CEPF), Beijing Environmental Protection Foundation (BEPF), China Wildlife Conservation Association (CWCA), Beijing Wildlife Conservation Association (BWCA), China Green Foundation (CGF), China Association of Environmental Protection Industry (CAEPI), Beijing Association of Environmental Protection Industry (BAEPA), Botanical Society of China (BSC), Chinese Society of Natural Resources (CSNR), Chinese Society for Environmental Sciences (CSES), Green Camp of University Students in China (GCUSC), China Green Student Forum (CGSF), Student Green Association of Tsinghua University (SGATU), PKU Green Life, Scientific Exploration and Outdoor Life Society of Beijing Forestry University (SENOL), Shanghai Teen Environmental Enthusiast Association (STEEA), and Center for Legal Assistance to Pollution Victims (CLAPV). These organizations lead environmental activities that provide Chinese society with many public benefits that are difficult for governments and companies to provide and promote environmental movements across China since the country first began reform and opening up. Local environmental organizations work on eight issues. (1) They hold promotional and educational activities to increase environmental awareness. Many NGOs hold awareness activities including various environmental initiatives and practical activities; televised and live lectures, trainings, speeches, and other educational activities about environmental awareness; and seminars, experience sharing opportunities, symposia, etc. (2) They promote and facilitate public participation in environmental protection. Governments and NGOs are increasingly cooperating on public participation in environmental protection in many Chinese cities, especially Beijing. (3) They fund environmental protection activities. International and overseas NGOs and foundations active in natural resource and environmental protection and Chinese foundations and NGOs related to environmental protection provide support through financial, equipment, technical, and other means for activities protecting natural resources and the environment. (4) They conduct projects protecting natural resources and the environment. Many NGOs are implementing various environmental projects, including wildlife and biodiversity protection, natural ecosystem maintenance and protection, tree planting, water purification, air pollution control and treatment, desertification prevention, soil erosion prevention in the upper reaches of the Yellow River, community-based environmental protection, garbage classification and separation, recycling, etc. (5) They research, develop, and popularize science and technology for environmental protection. Environmental protection societies, research institutes, and other NGOs employ many Chinese authorities and experts in academic fields related to the environment. Research and development of the related disciplines and technologies along with R&D results promote science and technology that protects the environment in China. (6) They manufacture and promote products that protect the environment and form industry alliances. Chambers of commerce, industry organizations, and other economic organizations active in environmental protection facilitate the development, production, distribution, and consumption of environmentally friendly products among other important activities. (7) They help environmental pollution victims. Pollution victims form a special disadvantaged group that receives attention from across society as pollution worsens. NGOs provide pollution victims assistance and legal advice. (8) They hold international exchange activities for environmental protection. The vast majority of environmental NGOs exchange information intentionally. They actively seek support from the international community for information, finance, equipment, technology, and in other fields. Environmental NGOs are further strengthening international cooperation for environmental protection by holding and attending international conferences and symposia, sending personnel for training, receiving visitors, and making visits. The Chinese government must follow and nurture NGOs, allow them to contribute to environmental protection, and encourage NGOs to participate in cooperation and exchanges for eco-city development projects. NGOs should continue advocating for an eco-civilization.

Increasing Public Participation in Eco-City Development The public produces and consumes in the city and builds and protects it, so that successful eco-cities outside China encourage the broadest possible public participation. Specific measures ensure broad public participation in the development and implementation of plans and schemes and in follow-up and monitoring. Many cities have benefited from such participation. Broad public participation is considered indispensable to the success of eco-cities outside China. Behind each objective in the local Agenda 21 developed by Seoul, South Korea, are specific fields and means for public participation to ensure project implementation. The government of Osaka, Japan, launches an annual public garbage collection event in September. It sends pamphlets to one million households that explain how to treat and recycle waste. The government of Montreal, Canada, uses T-shirts printed with public service ads, calendar cards, notebooks, and buses to call for public participation in waste recycling.

Eco-city development's massive scope requires broad public participation. A city must provide its citizens with environmental education on its way to becoming an eco-city in order to raise environmental awareness, create a sense of responsibility to protect the environment, increase resource and energy conservation and material recycling, promote a green and thrifty consumption pattern, and instill adherence to environmental laws. Public participation is initiated by governments in most cases and focuses on environmental education and monitoring but lacks pre-project participation. Governmental organization is more important that institutional support in boosting public participation. The Chinese public often discusses the environment but is not prone to act. China needs an established system for public participation in environmental protection.

The public can participate in environmental protection by learning and disseminating environmental information, making environmental decisions, and monitoring environmental policies. Environmental coverage and education must be enhanced to raise awareness of environmental rights and of the right to participate in public affairs and to spread eco-civilization ideology to individuals, families, and the entire society if public participation is to increase. Governments should hold more activities for public participation in environmental protection. Clear approaches to public participation in environmental protection will increase it. The public cares about and wants to participate in environmental protection, but surveys indicate people do not know how to engage specific environmental issues. Rules and regulations, such as the 2006 *Interim Measures for Public Participation in Environmental Impact Assessment*, help ensure the public's right to participate in environmental decisions. Public monitoring of environmental protection can stand improvement. The public is most concerned about environmental issues because its quality of life is directly intertwined with the environment. Public monitoring of environmental protection can improve by informing people of hotlines for reporting environmental issues. Promoting and improving environmental public interest litigation provides another avenue to involvement. Environmental public interest litigation can be any citizen, community group, or government agency that files a lawsuit to protect the public interest. It is necessary to systematically manage the growing public requirements for environmental rights. The eco-civilization is an entirely new form of human civilization based on reflections and adjustments to traditional civilizations, especially the industrial civilization. The eco-civilization is a major milestone for human civilization development.[12] It emphasizes human consciousness, self-discipline, interdependence, and mutual promotion and harmony between humanity and the natural environment. Eco-city development is integral to eco-civilization development and should respect nature; achieve and maintain harmony between human beings, nature, and society and between human beings among themselves; establish sustainable production and consumption patterns; and guide people toward sustainable and harmonious development. Eco-city development requires mobilization of positive factors; coordination of relationships between human beings, nature, and society and between human beings themselves; and planning and management based on ecological principles, in order to establish harmony for human, society, and nature.

2.3 Recommendations for Future Eco-City Development in China

China is shifting its attention from the rapid economic growth pursued over the past 30 years to enhancing public welfare. Urban development should make a similar shift away from economic growth focus and toward people. Many Chinese cities are still struggling to coordinate livability, economic growth, and productivity. How can Chinese cities develop into eco-cities? Four approaches present themselves. First, Chinese cities must go green to address urban economic growth and reduce carbon emissions. Second, Chinese cities must develop a circular economy to achieve cleaner production. Third, Chinese cities must promote low-carbon lifestyles to improve the living environment. Fourth, Chinese cities must advocate cultural livability to solve ecological problems within urban residential areas.

[12] Zhou [11].

2.3.1 Promoting Green Development

What Is Green Development? Green development encompasses a new, integrated pattern of economic, social, and environmental development that employs reasonable and low consumption, low emissions, increased ecological capital, and green innovation. Green development accumulates green wealth and welfare for human beings and achieves harmony between human beings and nature and among human beings.[13] Green development contrasts black development, which accepts pollution with plans for later treatment. Green development advocates symbiotic coexistence between human beings and nature and incorporates sustainable development. This includes sustainable development of the planet and humanity, striving to achieve green economic growth, accumulating green wealth in natural systems, increasing green welfare in social systems, and shifting a country's or region's ecological deficit toward ecological surplus. Going green is a universal solution to issues such as global warming and is the only path to promoting China's green rise.

How Does Green Development Relate to Eco-City Development? Green development guides urban development toward a green city and leaves profound and lasting effects on the development process. (1) Green development promotes green innovation during urban development. Eco-cities around the world have been the leaders in green innovation. Freiburg, Germany, harnesses solar power; Kathmandu, Nepal, has adopted green roof measures and limitations on building heights; Reykjavik, Iceland, is using hydrogen fuel instead of oil; and Toronto, Canada, has adopted a LED lighting system. (2) The concept develops green industries. Specific measures may include developing the forest industry to increase forest coverage; developing new energy, materials, and renewable resources; eliminating energy-intensive and polluting industries; and developing low-carbon, green, modern services to create a green production system. (3) The concept promotes green consumption in cities. Specific measures may include promoting green food and medicine; encouraging energy-efficient appliances and household products; promoting green travel such as walking, cycling, and public transport; etc. (4) The concept promotes increasing green wealth and welfare in cities. Plenty of sunshine, fresh air, clean water, rich green spaces, vegetation, and other elements constitute green wealth and welfare in a city, and all contribute to a healthy and happy life and work environment for ordinary people. Initiatives for green development can accumulate green wealth for the city and improve green welfare for local residents.

Measures for Green Development in Chinese Cities Ouyang Zhiyun et al. wrote an evaluation report on green development in Chinese cities that brought attention to the low investment in pollution treatment, infrastructure, and waste recycling and the successes of greening efforts, municipal solid waste (MSW) treatment, reduced water consumption per unit of GDP, high air quality, and industrial wastewater treatment. Cities with particularly high investment are directly under the central

[13] Hu [12].

government or provincial capitals, but output value from comprehensive waste recycling in these cities is not high. Cities with particularly high water consumption per unit of GDP and failing industrial wastewater discharge standards tend to be smaller. Cities in Fujian, Guangdong, Hainan, Hebei, Hubei, Jiangsu, Jiangxi, Shandong, Zhejiang, and other coastal provinces or provinces near the coast, as well as the four municipalities directly under the central government, are further developed in these fields when compared with cities in Gansu, Jilin, and Shanxi and northwestern provinces such Shaanxi.[14]

Qu Geping, the former head of the Chinese NPC Environmental and Resources Protection Committee, proposed six criteria and nine approaches to building green cities. The six criteria include reasonable plans and spatial arrangements, reliable urban infrastructure, effective pollution prevention and high environmental quality, clean energy use, a minimum percentage of natural landscape covering the city, and residents with strong environmental awareness. The nine approaches include developing reasonable urban plans; implementing an economic development pattern around a circular economy; building a fully functional urban environmental infrastructure; establishing a fast, convenient, and clean urban transportation system; creating an energy system led by clean energy sources; building beautiful, eco-friendly, residential areas with necessary services and high environmental quality; researching and developing an environmentally friendly technical support system; improving the sustainable development legal system; raising awareness around environmental protection and resource conservation across society; and promoting eco-friendly values and green consumption and urging public figures, especially government leaders, to accept sustainable development and meet environmental requirements in when making decisions and shaping consumption.[15] These criteria and approaches determine developmental goals for green cities and provide an ecological perspective to development and environmental protection efforts, which is key for overall green development in Chinese cities.

2.3.2 Developing a Circular Economy

What Is a Circular Economy? A circular economy is an economic growth pattern that efficiently uses resources, adheres to the 3Rs (reduce, reuse, recycle), enjoys low consumption and emissions, and is highly effective. Reduction happens as people use fewer raw materials and energy inputs produce and consume, conserving resources and reducing pollution at the first phase of economic activities. Reuse is realized as people use purchased articles, reuse wastes, and extend product and service lifecycles. Recycling requires waste discharge and emission minimization and conversion of used items into usable resources instead of disposing them as useless garbage. The circular economy establishes a "food chain" and "food web"

[14] Ouyang et al. [13].

[15] Guo et al. [14].

in the eco-economic system by simulating natural ecosystems, creating a producer-to-consumer-to-decomposer circulation path in the economic system, and using a symbiotic network to achieve closed-loop material recycling and multilevel energy utilization.

The circular economy is a new pattern of economic development that appeared as resource, environmental, and economic problems worsened. The circular economy organically combines industrialization, urbanization, and eco-friendly development, provides a path toward sustainable development, and offers a solution beneficial to the economy and the environment. It is significant in two dimensions. First, the circular economy marks a departure from the old and unsustainable economic system of mass production, consumption, and waste. It reconstructs the economic system to material and energy circulation in natural ecosystems, enabling a harmonious and cooperative relationship. Second, the circular economy integrates cleaner production with comprehensive waste recycling and respects ecological laws when using natural resources and environmental capacity. All raw materials and energy are used in a reasonable and exhaustive fashion in ongoing economic circulation, minimizing the environmental impact of economic activities and improving ecological growth. The change renders economic activities increasingly eco-friendly.

How Does the Circular Economy Relate to Eco-City Development? The circular economy satisfies eco-city requirements by achieving low input, high output, and low emissions production linking it closely to eco-city development. (1) Circular economy theory forms a basis for eco-cities, and developing a circular economy provides a new paradigm for eco-city development. A city can maximize economic and social development with minimal environmental and resource costs and can protect resources and the environment with minimal economic and social costs by following circular economic theory. Cities are not required to sacrifice the environment for development nor abandon development for protection. A city can maintain the finest environment and quality of life for local residents and at the same time ensure sustainable, healthy, and rapid socioeconomic development. The circular economy prompts the city to embark on a technology-led, resource-conserving, and eco-friendly modern development pattern. (2) Eco-cities result from circular economy development, ensuring their economic systems will follow circular economic principles. The current open-loop economic process moving from resources to products to waste should be closed into a circular economic process going from resources to products to resources using the producer–consumer–decomposer mechanism found in natural ecosystems, thus reducing resource consumption and recycling wastes into resources. (3) The circular economy is a necessary method and pattern of eco-city construction. Developing a circular economy requires organizing economic activities under ecological and economic laws. The circular economy requires a city to better utilize nonrenewable resources, replace nonrenewable resources with renewable alternatives to the greatest extent possible, and establish a green consumption system, thus achieving maximum reduction to resource and the environmental damage caused by human consumption and sustainable consumption. The circular economy also requires a green GDP accounting system and a focus on resource conservation and environmental protection.

Measures for Developing a Circular Economy in Chinese Eco-Cities Resource use reveals three reasons for serious environmental problems across China including rapidly growing resource consumption, inefficient resource use, and a low recycling rate. China must address these problems by developing circular economy if the trend is to shift. China has formed a unique circular economy development pattern referred to as the "3 + 1" pattern, which includes small-, medium-, and large-scale circulation plus waste treatment and reproduction. Small-scale circulation occurs at the enterprise level. Large and average enterprises are selected for pilot eco-industrial projects that implement green product design, cleaner production, and other measures with a goal of eco-efficiency, reducing the consumption of materials and energy for products and services and minimizing pollution. Medium-sized circulation occurs at the regional level. Inter-enterprise integration of materials, energies, and information creates a symbiotic relationship among enterprises and develops EIPs in accordance with industrial ecology. Large-scale circulation occurs at the national level. It focuses on building cities and provinces with circular economic systems. Liaoning Province has begun building circular economy demonstration zones focused on transforming old industrial centers. Guiyang City in Guizhou Province is developing into a circular economy city focused on comprehensive use of local resources unique to the area to create a new industrial pattern. Waste and used resources are treated, disposed of, and reused in production as per the relevant regulation for recycling waste and used resources across society. Planning, design, and construction of circulation systems are at the center of circular economic development needed for eco-cities. Urban circulation planning and design uses clean and renewable energies, improves resource utilization, and generates no wastes or garbage. Urban circulation planning and design should rely on local natural resources and with a consideration of socioeconomic and technological development. The focus should remain on assessing the carrying capacity of ecosystems, natural landscapes, and the hydrographic environment. Relevant organizations should integrate the city's economic system with the natural environment. Urban circulation planning and design include circulation design for production and industry. Product circulation planning and design ensure that product design, production, and sales are all green. Product designers should consider quality, functionality, price, aesthetics, and other factors that routinely satisfy human consumption utility as well as the environmental impact of a product throughout its lifecycle to ensure green product designs. Such considerations minimize the product's environmental impact without reducing its consumption utility. Green manufacturing and production including resource conservation and rational utilization can be achieved through tight control over resources and the entire production process. Wind power, solar power, bioenergy, and other clean renewable energies should be used widely, and energy efficiency can increase through cogeneration, energy sharing, and multilevel energy use. Nontoxic, renewable, and environmentally degradable raw materials should replace toxic, nonrenewable, and nonbiodegradable materials to reduce environmental impact. Sales prices in the product marketing stage should include profits from marketing events while reflecting environmental damage caused through product consumption to meet corporate interests, consumer demand, and environmental interests. Extra attention should be paid to green product packaging, ensuring reusable package materials. Waste caused by packaging, production, and consumption should be returned to the manufacturing process and recycling energy and materials. Circulation design should focus on building an eco-friendly industry chain. The entire process results in resource–product–resource circulation within the enterprise, ensuring an eco-friendly industry chain for the production process and an ecosystem network between production and consumption that covers all three activity sectors. The core of all circulation rests in circulation within enterprises, which allows circulation throughout the city. The Law of the Human Body theory sees a city like a human body and business communities consisting of interrelated downstream and upstream enterprises as its organs, which perform specific complex functions that circulate materials, energy, and information through the city's arteries and veins, thus connecting the city with the country, its suburbs, and other cities.

2.3.3 Advocating a Low-Carbon Life

What Is a Low-Carbon Life? The world is shifting toward a low-carbon economy, and countries are competing for leadership in this trend. A low-carbon economy seeks the highest possible output that still reduces greenhouse gas emissions. It is one of fundamental tools for humanity in dealing with climate change and among the most recent economic trends around the world in the twenty-first century. Greenhouse gases in the atmosphere absorb and emit thermal infrared radiation. Human activities alter levels of greenhouse gases such as carbon dioxide, methane, nitrous chloride, hydrofluorocarbons, perfluorocarbons, and sulfur hexafluoride. Carbon dioxide has the greatest impact on climate change. A low-carbon economy reduces carbon content in the production or consumption process to yield lower carbon dioxide emissions and cleaner air. A low-carbon economy is an economic development pattern based on low energy consumption, pollution, and emissions that marks a significant milestone for human society following the agricultural and industrial civilizations. A low-carbon economy works toward efficient energy use, clean energy development, and green GDP. Energy and emission reduction technology innovation and development in industry form the core of a low-carbon economy along with fundamental changes to the concept of human survival and development.

All economies are facing significant pressure to reduce carbon as resources are exhausted, energy prices soar, environmental pollution worsens, and the global climate loses balance. The traditional high-carbon development pattern that relies on the massive consumption of fossil fuels is no longer sustainable. Future development must seek low-carbon solutions offering strategic advantages to lead new energy and low-carbon technology development, as they will lead the future green industrial revolution and a new round of global economic growth. China has been attentive to energy conservation and environmental protection since it began reform and opening up and has increased interest since the 1990s. China's economic growth pattern remains inefficient, engaging in economic development largely at the expense of the environment, because of the traditional GDP-focused evaluation system. The Chinese economy has arrived at a crossroads in the twenty-first century after a burst of rapid economic growth and new development in the industrial and chemical sector. China must embark on a low-carbon transition focused on energy conservation and emission reduction to transform the economic development pattern following the Scientific Outlook on Development.

How Does a Low-Carbon Life Relate to Eco-City Development? Cities around the world house more than half of the world's population and account for 75 % of total greenhouse gas emissions. Cities have a massive demand for resources and huge carbon emissions that go far beyond their carrying capacities and have negatively affected their continued development and their role in local and global environments and economies. The high density urban environment and wide range of activities available in cities that create their huge resource demand do not excuse cities from developing a low-carbon economy. Cities can use their large and dense populations and variety of commercial and industrial activities to more effectively control resource and energy consumption per capita than low density areas. Large cities wielding regional power have advantages for developing a low-carbon economy than their political and institutional frameworks. Low-carbon cities are crucial for leveraging these advantages. A low-carbon economy results from efforts to reduce greenhouse gas emissions but is also a broader revolution in economic development, energy consumption, and human lifestyles. A low-carbon economy will transform the entirety of the modern industrial civilization based on fossil fuels for energy sources into an eco-economy and eco-civilization. It increases energy efficiency and improves the clean energy source mix. Energy technology and institution innovations such as minimizing high-carbon energy sources such as coal and oil in economic activities and furthering low energy consumption and low pollution in the economy are at the center of the low-carbon economy. A low-carbon economy is the only way to achieve sustainable urban development. Implementing these economic values in the city by furthering low-carbon production and consumption, conserving resources, greening society, and promoting a healthy and sustainable energy ecosystem creates a low-carbon city. A low-carbon city begins foremost with developing low-carbon energies but can also include cleaner production, recycling, and sustainable development as an end objective.

Measures for a Low-Carbon Life in Eco-Cities China is industrializing and urbanizing rapidly. It has entered the second half of the intermediate industrialization stage. Heavy industries and chemical industries are developing rapidly leading energy-intensive industries to an increasing share of the economy and to increased energy consumption. China's total greenhouse gas emissions increased sevenfold between 1970 and 2007. China overtook the United States as the world's largest carbon dioxide emitter in 2007. Coal-fired power plants in China's industrial sector account for over 84 % of total carbon emissions. The industrial sector has higher emissions intensity per unit of GDP than agriculture and services. China could reach 9.5 billion tons of carbon emissions in 2020, which would represent 27.8 % of the world's total carbon emissions, according to forecasts by the International Energy Agency (IEA). In 2030, that amount will go up to 11.6 billion tons of carbon or about 30 % of the world's total carbon emissions. China's urbanization process has developed rapidly alongside economic development. China's urbanization level is only at 46.6 % and will only reach 75 % in 30 years if growth remains at one percent per year. Such a trend means China's urban population will increase by 450 million by the end of 2040, equivalent to 1.5 times the US population and equivalent to the population of the EU. Rapid urbanization has significantly increased the scope of urban housing, transportation, and other infrastructure, as well as energy and resource consumption, multiplying the city's greenhouse gas emissions. This presents China with a complex set of domestic and international factors in addressing sustainable development and environmental protection. China must avoid energy demand and greenhouse gas emissions locking into economic development if it is to continue urban development. It must quickly transform the economic development pattern, using low-carbon, energy-conserving, and environmentally friendly technologies, developing low-carbon cities and eco-cities, and promoting green low-carbon development in urbanization.

The NDRC, the MOHURD, the Tianjin Municipal Government, and the China Center for International Economic Exchanges (CCIEE) held the Second China (Binhai Tianjin) International Eco-city Forum & Expo (Binhai Forum) on September 23–24, 2011. NDRC Deputy Director Xie Zhenhua attended the opening ceremony and delivered a keynote speech titled "Promote opportunities through green low-carbon development for low-carbon cities" at the main forum, proposing a number of measures for low-carbon cities. (1) He proposed optimizing the industry mix; developing the service sector and emerging strategic industries such as energy conservation, environmental protection, and new energies; and preventing energy-intensive, high-emission industries from rapid growth. Enhancing binding targets for energy conservation, environmental protection, land use, and safety, eliminating obsolete production equipment and technologies, and promoting the transformation of traditional industries for an energy-efficient, low-carbon industrial system further aid optimization. (2) He recommended optimizing the energy mix, controlling total energy consumption, strengthening cleaner production and coal use, and developing clean, efficient, high-capacity coal-fired generation units. Further steps include increasing oil and gas exploration and development, developing unconventional oil and gas resources, and developing nonfossil energy sources. Safe nuclear power, eco-friendly hydropower, wind power, solar power, biomass, geothermal energy, and other renewable energy sources can be developed based on local conditions. (3) He proposed strengthened energy conservation and low-carbon development management, energy efficiency benchmarking activities, and energy conservation and emission reduction projects in buildings, transportation, and other sectors. The "10,000 low-carbon businesses" initiative can help the industrial sector, the "green building" initiative can help the construction industry, and the "1,000 low-carbon transportation companies" initiative can help the transportation sector. Energy conservation and emission reduction can be better applied in agriculture, rural areas, commerce, households, and public institutions, and low-carbon projects are needed across society, including building low-carbon governments, campuses, and avenues. (4) He recommended accelerating research, development, and application of low-carbon technologies, strengthening basic scientific research on greenhouse gas emissions control, and researching and developing of cutting-edge technologies critical to reducing carbon. Governments should offer incentives for the broad application of energy conservation and carbon reduction technologies and equipment and support energy-efficient motors, semiconductor lighting, low-frequency waste heat utilization, and other important low-carbon technologies to advance their production and application.[16]

2.3.4 Advocating Cultural Livability

What Is Cultural Livability? The concept of livability first appeared in the garden city movement in the United Kingdom in the nineteenth century. Sir Ebenezer Howard presented the garden city theory in his book *Tomorrow: a Peaceful Path to Real Reform* in an effort to solve urban problems at that time by building comfortable, convenient, and beautiful garden cities (Howard, 1898). Urban planning has focused on creating comfortable and pleasant environments since World War II. In his book *Amenity and Urban Planning*, David L. Smith used examples from the second half of the nineteenth century to advocate for a pleasant environment using the amenity concept. Amenities cover public health and pollution, a comfortable living environment, historic buildings, and a beautiful natural environment. The WHO summarized the basic living requirements for human beings in 1961 as safety, health, convenience, and comfort to form the concept of a living environment. The second United Nations Conference on Human Settlements (Habitat II), held by the UN-Habitat in 1996, adopted the Habitat Agenda and declared a city a human settlement suitable for living. The conference established three fundamental criteria for livability including spatial, social, and environmental characteristics and qualities of an urban community, and it recognized democracy, consultation, public participation, and system development as external support for forming livable cities and towns. The concept of livability has since spread to other fields.

The famous Chinese scholar Wu Liangyong developed a living environment theory dividing it into five levels: global, regional, urban, community, and building. He proposed five principles for solving living environment problems, including recognizing the ecological predicament and raising ecological awareness; achieving positive interactions between living environment and economic development; developing science and technology and promoting economic development, social

[16] Xie [15].

prosperity, and the living environment through diverse methods and techniques; caring for the general public and respecting the overall public interests; and combining science with art.[17] Luo Yameng analyzed Chinese traditions and human settlement history to conclude that livability in China has evolved from seeking an affordable, pleasing, and large place to live with a positive, comfortable, and healthy living environment to become a search for a positive, healthy, peaceful, and safe place to settle permanently.[18] China promulgated the Scientific Evaluation Criteria for Livable Cities in 2007, realizing nearly 100 indicators spread over six categories: social civilization, economic prosperity, environmental beauty, resource-carrying capacity, living convenience, and public safety. Gu Wenxuan believes that a livable city and evaluation criteria for livable cities incorporate a wide range of disciplines and that efforts must look beyond comfortable living conditions and environment to thriving cultural and social environments with moral behavior, stable social order, extensive social welfare, full employment, etc.[19] For anything in China's current state of social development to be considered livable, it must be oriented around people. Livability depends predominantly on whether the needs of residents are met and whether they are satisfied with the city's living environment. A community with a high livability index and high life satisfaction will satisfy residents' need and is therefore a truly livable community. Building livable cities has gained prominence among twenty-first century cities around the world. Beijing first announced livability goals in 2005. A livable city is suitable for work and life using coordinated economic, social, cultural, and environmental development, a good living environment, and meeting the physical and cultural needs of residents. A livable city coordinates cultural and natural environments, continuous economic prosperity, social harmony and stability, a vibrant cultural atmosphere, and a full range of amenities. Shanghai, Guangzhou, Chengdu, Hangzhou, Dalian, Qingdao, etc. have all announced livability goals, greatly popularizing livability in Chinese urban development. Livable means suitable for human habitation so that livability efforts require orientation toward people. A livable city views a city as a collection of people's lives and aims to improve the overall quality of life and to achieve harmony between humanity and the natural environment. A city's livability should be present in its economy, culture, society, and environment. A livable city should meet comprehensive livability requirements including a prosperous urban economy, a pleasant environment, excellent public services, a harmonious cultural atmosphere, etc. We define a city's livability by the quality of urban life available to its residents. A livable city must meet a number of requirements. (1) It must be livable as a whole. A livable city must extend to economic, cultural, social, and environmental domains. (2) A livable city should have a prosperous urban economy with a stable and healthy economic environment, a rational industry mix, developed and modern services, adequate and stable employment opportunities, and job providing high disposable income necessary for a livable city. (3) A livable city should have a flourishing envi-

[17] Jin [16].

[18] Luo [17].

[19] Gu [18].

ronment providing fresh air, clean water, lush greenery, and beautiful natural landscapes, all necessary for ecological livability. (4) Comprehensive public services, including urban public facilities, transportation, housing, safety and security, disaster reduction, employment, medical care, education, welfare, etc., constitute livable service in a city. (5) A livable city should have a harmonious cultural atmosphere, including social order, morality, culture, entertainment, etc., to provide cultural livability.[20] Li Liping and Guo Baohua, researchers at the Institute of Regional Economy and Urban Management and Renmin University of China, codeveloped an indicator system for evaluating livable cities that offer a set of Level 1 indicators covering economy development, social harmony, cultural richness, living comfort, landscape, and public safety. (1) A livable city should enjoy continuous economic prosperity. A city centralizes economic organization, management, and coordination. Various economic actors and nonagricultural activities find their home in the city. A strong economic foundation, an advanced industry mix, and growth potential allows a city to provide the adequate job opportunities and high incomes desired by urban residents in a livable city, as well as resources for infrastructure. (2) A livable city should harbor social harmony and stability. A city with political stability, security, peace among ethnic groups, harmony among classes, attractive communities, and coordinated urban and rural development provides residents the opportunity to live and work contently, to fully enjoy modern urban life, and to regard the city as their physical and spiritual home. (3) A livable city should offer rich culture. Rich culture can include rich historical and cultural heritage, a full range of cultural facilities, frequent cultural activities, and a supportive cultural atmosphere. Rich culture in a city offers a rallying point for ideology, education, technology, and cultural experiences, providing education unique to an urban environment and improving the urban population. (4) A livable city should offer comfort and convenience in all domains but especially in residence. A livable city should provide a full range of amenities, residential buildings that meet health requirements, convenient transportation that include a developed public transportation network, as well as sufficient public goods and services. Examples include sufficient and quality education, medical care, sanitation services, and maintenance of healthy ecosystems, good air and water quality, quiet and clean residential areas, and a large per capita green space area. (5) A livable city should provide pleasant and beautiful landscapes. Beautiful and pleasant landscapes are a basic requirement for the cultural and natural mix present in urban development. Cultural and natural landscapes must be coordinated in the city to allow a good mix of cultural landscapes such as roads, buildings, squares, and parks that address human needs and development to improve residents' moods. (6) A livable city should be a place where the public feel safe. A city can provide public safety and security by protecting its residents and their property against natural disasters such as earthquakes, floods, storms, and infectious disease epidemics and against disasters caused by humans. Residents can live and work

[20] Tong [19].

contently only if they feel they are safe, making public safety and security an absolute necessity for a livable city.[21]

Measures for Building Culturally Livable Eco-Cities in China Urban development aims to improve people's lives. China should take the following measures to develop culturally livable eco-cities:

(1) Promote public participation in eco-city construction. Public participation in major urban planning decisions, management, and development is a key part of urban development. Public participation promotes urban sustainable development by helping managers understand the public's needs and the way various plans and decisions will impact them, guiding managers toward innovation that will satisfy the public's wishes, and producing plans that pursue urban diversity and energize the city. Livable cities outside of China have demonstrated that public participation guides livable city development. Chinese cities may refer to the experience foreign cities have in integrating public participation to increase public participation in policy making and coordinate and balance the relationships among all stakeholders through techniques like phased implementation, tracking and feedback, and regular evaluation and revision.
(2) Promote economic development. A livable city should enjoy a developed economy. Economic development is the foundation of social progress. Sustainable economic development is the only means to solving a series of urban problems such as poverty, environmental pollution, and underemployment and to creating a tangible and vibrant living environment for residents, which promotes the creation of an intangible living environment. A livable city must have strong economic growth potential to ensure sustainable economic development, improve residents' living standards, and provide a positive living environment for residents. Chinese cities may attract investment from more international companies because of their low level of economic development through flexible policies and can access local resources to develop related industries and unique city qualities while furthering industrial restructuring and enhancing advanced technology research and development, which can be used to revitalize obsolete enterprises. These measures will contribute to the sustainable development of livable cities.
(3) Create beautiful spaces. A beautiful and pleasant environment is the most intuitive symbol of a livable city. Typical livable cities outside of China have almost all realized many benefits of eco-cities and developed specific skills geared toward creating beautiful spaces. Most Chinese cities have clearly stated their intention to conduct eco-friendly urban development and create a pleasant living environment and pleasant living and working spaces, but they are still new and inexperienced in environmental development leaving room to learn. Many Chinese cities are still learning how to fully profit from natural ecosystem resources, effectively organize natural landscapes, design green spaces, create a pleasant urban atmosphere, and build various activity spaces.

[21] Li et al. [20].

(4) Create livable residential spaces. The housing problem encompasses fundamental household interests and affects social fairness and stability. A livable city must address these extensive housing concerns. Livability implies that every person who needs to live in a city permanently has access to a decent residential space. Singapore is widely recognized for its competence in this regard. The government provides housing and runs an agency specializing in related policies. Housing prices are determined according to most residents' purchasing power, and incentives are available to ensure residents can afford a house. A livable city should ensure local residents can afford decent apartments or houses and rest permanently. Community environment must be incorporated into urban development. Relevant organizations should focus on improving community safety, greening, health, culture, and other projects. It is necessary to enhance community facilities, provide aid to disadvantaged groups, create a positive cultural atmosphere in the community, and help residents improve their behavior. Efforts should be made to create a quiet and clean living environment where residents have a sense of belonging and feel safe.
(5) Establish an efficient transportation system. Transport development is a necessary priority to ensure convenient travel around a livable city. Convenient and comfortable transportation is a primary factor leading many people to stay in livable cities outside of China. The number of vehicles in rapidly urbanizing Chinese cities has been rising sharply, creating consistent transportation issues that affect urban efficiency, socioeconomic development, and public health. A livable city should have reliable and efficient transportation. It should build reasonable transportation infrastructures, improve traffic management, and develop public transportation options to achieve this goal. A people-oriented, pleasant, and complete pedestrian network will further facilitate citizens' recreation and travel. A livable city can reduce unnecessary motor vehicle traffic by coordinating the flow of passengers and cargo and urban land use.
(6) Strengthen urban safety and security. Cities inside and outside of China have historically always prioritized safety and security. A safe city can maintain a dynamic, stable, balanced, and coordinated environment, ecosystem, socioeconomic development, culture, health, resource supply, government performance, etc. It can protect against natural disasters and socioeconomic anomalies or emergencies. Urban safety and security increasingly concerns governments and the public across China as urbanization increases and complicated interests and conflicts emerge between human populations and resources and between the environment and development amid the social transition. Chinese cities do not properly address safety and security measures nor do they have a comprehensive urban safety and security system as part of eco-city development. Some cities have begun taking steps toward safety and security but remain in the early stages. A safe city provides citizens with security surrounding the environment, food, the society, production, the economy, and culture.
(7) Preserve the city's cultural characteristics. A livable city celebrates its own character and qualities rather than imitating other cities. The culture of a city best reflects its character. A city with unique character and qualities is competi-

tive, has great growth potential, and is likely to develop into a livable city because it resonates with the regional culture. Chinese cities currently lack many of these key characteristics. Most cities resemble concrete jungles. A livable city's cultural characteristic goes beyond a continuation of the traditional culture but does not amount to arrogance. A livable city's distinctive cultural environment can flourish when the traditional culture is mixed with modern elements. A livable city preserves its historical and cultural heritage, tangible and intangible to enrich local citizens' cultural life. It integrates a diversity of cultural elements, drawing from different times and geographic areas, and concentrates native cultural elements in the city.[22]

Building eco-cities is a basis for building a beautiful China. The report to the 18th CPC National Congress states that "promoting ecological progress is a long-term task of vital importance to the people's well-being and China's future. Faced with increasing resource constraints, severe environmental pollution, and a deteriorating ecosystem, we must raise our ecological awareness and respect, accommodate, and protect nature. We must prioritize ecological progress and incorporate it into all aspects of advancing economic, political, cultural, and social progress, work hard to build a beautiful country, and achieve lasting and sustainable development of the Chinese nation." This will fuel eco-city development across China, usher more Chinese cities toward eco-city development, and further enrich theories regarding eco-city development.

References

1. Li X, Liu Y (2011) Eco-cities: opportunities and challenges. China Construction News, 17 June 2011
2. Fan G et al (2011) Low carbon city in action: policy and practice. China Economic Press, Beijing, p 27
3. China Society for Urban Studies (2012) Report on low-carbon eco-cities in China 2012. China Architecture & Building Press, Beijing, p 33
4. Tian H et al (2012) Development status and prospect of green buildings in China. Build Sci (4)
5. Richard R (2010a) EcoCities: rebuilding cities in balance with nature. Social Sciences Academic Press, Beijing, p 3
6. Richard R (2010b) EcoCities: rebuilding cities in balance with nature. Social Sciences Academic Press, Beijing, p 167
7. Che W (2012) Ten questions about sewers: the 10th report. People's Daily Online, 22 Oct 2012
8. Wu L (2001) Introduction to sciences of human settlements. China Architecture & Building Press, Beijing, p 46
9. Zhu G (2008) "Implement the scientific outlook on development and promote the building of eco-districts/cities – a speech at a site meeting about eco-district/city development nationwide"; the Department of Nature and Ecology Conservation, the MEP. In: Eco-province development: theories and practices. China Environmental Science Press, Beijing, p 68

[22] Jiang et al. [21].

10. Ye J et al (2009) On building a cultural ecosystem in the process of eco-city development. Changbai J (2)
11. Zhou S (2008) "Development toward eco-civilization: the concept of eco-civilization and the significance of development toward it," the Department of Nature and Ecology Conservation, the MEP. In: Eco-province development: theories and practices. China Environmental Science Press, Beijing, p 23
12. Hu A (2012) China: innovative green development. China Renmin University Press, Beijing, p 33
13. Ouyang Z et al (2009) On the green development of Chinese cities. China Popul Resour Environ 19(5)
14. Guo H et al (2003) On green city design in China. Acta Agriculturae Universitatis Jiangxiensis (Humanit Soc Sci) 2(4)
15. Xie Z (2011) Promote opportunities through green low-carbon development for low-carbon cities. Recycl Resour Circu Econy (10)
16. Jin W (2011) On Wu Liangyong's living environment science and methodology. J Urban Reg Plann (1)
17. Luo Y (2006) What makes a livable city. Sci Deci Mak (12)
18. Gu W (2005) Building harmonious cities is an important issue for scientific research on cities. Urban Stud (12)
19. Tong W (2007) On the livability strategy for urban development. Thesis for a master's degree from the Southwestern University of Finance and Economics
20. Li L et al (2006) On livable cities. Urban Stud (2)
21. Jiang Y et al (2009) Livable cities outside China: practices and inspiration. Urban Plann Overs (4)

Part II
The Development of Eco-cities in China: Practices

Chapter 3
Environmentally Friendly Cities

Guohua Chang

3.1 Environmentally Friendly City: Concept and Definition

3.1.1 Concept

The following points outline the definition of an environmentally friendly city: (1) An environmentally friendly city centers on harmony between humanity and nature and adopts a holistic perspective incorporating all aspects of economic development, social behavior, the political system, culture, science, and technology in an organically unified development framework that will root out environmental problems during urban development.

Environmental problems may seem like a by-product of socioeconomic activities, but result more directly from conflicts between humanity and nature and between economic and environmental benefits. These problems arise from industrialization and the inappropriate economic, political, cultural, and technical patterns and systems that it instills. Environmental problems are multifaceted, covering many dimensions and factors and developing nonlinear and complex ways in nature, the economy, society, politics, technology, and culture.[1] An unscientific economic growth pattern driven by growing human demand has led to uncoordinated development of nature and humanity and many confrontations between them. Former Chinese President Hu Jintao remarked that many indicators suggest that disharmony between humanity and nature affects the relationships among humans and within society at the central government's meeting on population, resources, and the environment. Harmony among humans and within society will be difficult to achieve

[1] Ren et al. [1].

G. Chang (✉)
Lanzhou City University, Lanzhou, China
e-mail: 494164311@qq.com

J. Li, T. Yang (eds.), *China's Eco-city Construction*, Research Series on the Chinese Dream and China's Development Path,
DOI 10.1007/978-3-662-48153-0_3

if the environment is severely damaged and if working and living environments are deteriorating, especially due to an already tight supply of resources and energy and severe conflicts between economic development and resources and energy. Human-induced ecosystem damage has led to a shortage of natural resources, ecological degradation, and environmental pollution. Environmental issues extend beyond nature to social factors such as socioeconomic development. They spread beyond domestic concern to global interest, sometimes causing turmoil around the world.[2] Humans must adopt a holistic approach to incorporate all aspects of economic development, social behavior, political systems, culture, science, and technology into an organically unified development framework to root out urban environmental problems and redefine the urban development to focus on environmentally friendly approaches. (2) An environmentally friendly city establishes positive interactions between the urban and natural environments and builds harmony between humans and the natural environment. A healthy natural environment promotes comprehensive and coordinated urban sustainable development. Environmental problems in the last century have resulted from humanity's failure to correctly understand and treat their relationship with nature.[3] Humanity views itself as holding a dominant position in its relationship with nature. This dominant position should not imply humanity ruling over nature or acting as its master, which would cause confrontation and conflicts between humanity and nature and create an ecological imbalance and environmental degradation, threatening human society's sustainable existence and development. Humanity's dominance should result in motivation to action. Environmentally friendly urban development relies on correctly understanding the relationship between humanity and nature before adopting production, living, and consumption patterns that favor environmental protection and that maintain ecological harmony. Ensuring this understanding helps maintain human benefits and promote human development. Establishing harmony between humanity and the natural environment is the only way to achieve comprehensive and healthy development for humanity and for the natural environment. Environmentally friendly cities establish a positive interaction between urban and ecological environment in this way. Human activities in the natural environment are refined using social activities. The social activities must be friendly to the environment for this strategy to be effective. Governments, businesses, communities, schools, hospitals, household, etc. should each craft their activities to help ensure their city is environmentally friendly.[4]

(3) Environmentally friendly cities emphasize production, living, and consumption patterns respectful of the environment's carrying capacity and focus on socioeconomic activities' impact on the natural environment. The environment is an important element factoring in regional economic development and determines the regional environmental quality and the size of the regional economy. A combination of a good environment and sustainable use of natural resources ensures a

[2] Song [2].

[3] Fang [3].

[4] Zhang [4].

continuous supply of resources to support urban economic development. This combination is an important foundation for achieving sustainable urban development and to allow natural environmental renewal. Theories regarding the environmental carrying capacity stipulate that any natural or socioeconomic environment can only continue providing natural or socioeconomic resources within certain limits. Exceeding the environment's carrying capacity worsens environmental quality and exhausts resources, impacting human activities and health.[5] (4) Environmentally friendly cities should reflect their respective characteristics. Cities differ in terms of natural environment, resource endowment, economic development level, and culture leading to differences in natural conditions for development, constraints, and environmental problems. Cities should steer away from a prepackaged patterns or frameworks for environmentally friendly cities and instead carefully analyze their local environmental problems and their own conditions and characteristics to find their respective approaches to environmentally friendly development. Environmental friendliness should be sought after through differing means.

(5) Building an environmentally friendly city is a dynamic development process executed through various phases and undergoing continuous improvement. It is currently required that a city minimize the load of its economic activities and their impact on the environment before it brings these measures within the resource supply limits and self-purification capability of the environment, which allows a virtuous socioeconomic development and environmental cycle.[6]

3.1.2 Basic Characteristics of an Environmentally Friendly City

An environmentally friendly city is defined by the following characteristics: (1) Harmony. An environmentally friendly city nurtures harmony in human and natural relationships and more importantly in interpersonal relationships to provide an amiable atmosphere in nature, society, economy, institutions, the cultural environment, and the coordination between these aspects. (2) Sustainability. An environmentally friendly city develops sustainably, considers spatial-temporal factors, and rationally allocates resources to prevent exceeding the environment's carrying capacity, thereby satisfying both the current and the future generations' developmental and environmental needs. It ensures comprehensive, coordinated, and sustainable development through these methods.[7]

(3) Integrity. An environmentally friendly city pursues a beautiful environment and considers the overall effectiveness of its social, economic, and environmental

---

[5] Zhang and Zheng [5].

[6] Ren et al. [1].

[7] Su and Yi [6].

systems. It values coordination between economic development and the natural environment, improving the quality of human life. It seeks development with social, economic, and environmental system coordination. (4) Regionality. An environmentally friendly city is a regional concept in that each city has unique resources and conditions. Developing an environmentally friendly city boosts regional balance and specialization.

3.2 Creating Assessment Metrics for Environmentally Friendly Cities

3.2.1 Design Principles

The academic world has developed a number of principles for choosing indicators for sustainable cities, eco-cities, and environmentally friendly cities, requiring the indicators to be scientific, operable, comparable, reliable, representative, coupled along space and time, layered, and dynamic. Further research using these principles on assessment metrics for environmentally friendly cities has yielded the following guidelines:

Layered and Systematic Combination An environmentally friendly city is complex, open, and enormous, but can be divided into several subsystems. It is often difficult for humans to simultaneously consider all the details of complex issues; therefore, it is useful to decompose them into multiple layers. The hierarchical method sorts these issues from examinations that are more coarse to finer examinations, from outside to inside, and from the whole to parts. The sorting systematizes the process. Assessment metrics require the architecture of the system and indicators classified based on layers to clarify the metric structure and ensure they are clear and easy to use. Assessment metrics for environmentally friendly cities in this book are divided into three layers. The first layer is the target layer, the second layer covers subsystems of the assessment metrics, and the third layer includes specific indicators that constitute the assessment metrics.

Comprehensive and Representative Combination Assessing an environmentally friendly city involves all aspects of the city, which requires assessment metrics and mathematical models that fully reflect as many aspects of the city as possible and that accurately describe its current state and level of development, but which do not provide indicators that overlap. A representative and typical indicator set that avoids indicators that provide similar or identical information or indicators that can be derived from a combination of other indicators and contains as few indicators as possible will best present information on the main aspects of the assessment target. If the set incorporates too few indicators, it cannot adequately reflect the main features of the object, nor can it incorporate too many without generating excessive information.

Scientific and Feasible Combination The metrics should scientifically and accurately reflect the progress toward an environmentally friendly city. Each indicator should have a clear physical meaning, and methods for selecting data, calculating, and synthesizing results must all be based on recognized scientific theories. Estimations should rely on statistical methods that are scientific and adhere to current norms to ensure truthful and objective assessment results. Domestically or internationally accepted statistical indicators with reliable data sources are required. These indicators should be readily available with proven approaches to gathering statistics avoiding the need for surveys. Indicators for environmentally friendly cities should be simple, clear, and easy to understand and should consider data access difficulties, operability, and reliability. Existing statistics will provide the best results. The indicators should be in consideration with China's current level of economic development. Their methodology and human and material resources should match China's current productivity and consider the technical capacity of technical departments.

Comparable and Universal Combination The assessment metrics should allow for dynamic and side-by-side comparability. Dynamic comparability allows comparing a city to itself at different times in the process of developing into an environmentally friendly city, demonstrating the extent to which a city has achieved the title of environmentally friendly. Side-by-side comparability allows different cities to be compared at the same time based on their overall indicator values. It is used to describe the interregional imbalance of environmentally friendly urban development. The selected indicators should statistically align with previous assessment indicators and with indicators used in other regions of China and in other countries as much as possible to allow for comparisons. They should meet the requirements of China's existing statistical system in order to facilitate temporal and spatial comparisons. The selected indicators must be easy to compare with the indicators used in the past and those that will be used in the future, as well as with the corresponding indicators used in other regions. This ensures that an environmentally friendly city will be fully and correctly evaluated and that major factors affecting its development can be identified. Differences in economic development, industry mix, culture, environment, and resources between regions must be considered to choose comprehensive and universal indicators that will ensure that assessment results are geographically comparable.

Stable and Dynamic Combination Building an environmentally friendly city is a long-term, continuous, and dynamic process that requires assessment metrics that adequately reflect dynamic changes in the city. Timely adjustments and improvements to the assessment metrics and mathematical models should be made to reflect progress in constructing an environmentally friendly society, thereby dynamically reflecting changes in urban development. The assessment metrics and the mathematical models should remain relatively stable over a certain period and avoid frequent changes once established. This ensures effective comparison and analysis of the system's development process.

Forward Thinking with Orientation Combination Designing assessment metrics and mathematical models for environmentally friendly cities requires consideration of relevant trends and an incorporation of advanced indicators that anticipate future changes. These indicators reflect the past and present states and also incorporate future environmentally friendly urban development trends. The selected indicators should provide support and guidance for policymakers, the general public, and all organizations across society to encourage resource conservation, environmentally friendly behavior, and work toward the direction dictated by the perceptive indicators, allowing for a steady increase in environmental friendliness. These principles constitute an organic whole, interacting with and influencing each other, so that they should never be simply separated without considering their coordinated relationships in assessing an environmentally friendly city. These principles guarantee proper analysis and evaluation of environmentally friendly cities.[8]

3.2.2 Creating Assessment Metrics for Environmentally Friendly Cities

Composition Assessment metrics for environmentally friendly cities remain relatively unexplored around the world and public research reports are still rare. Creating assessment metrics for environmentally friendly cities must involve selecting from social, economic, environmental, and other factors based on the available understanding of the definition, composition, and characteristics of environmentally friendly cities. Assessment metrics for environmentally friendly cities should fully reflect the city's economic development, social progress, and environmental quality. Existing indicators of sustainable development and environmentally friendly cities serve as a reference along with a series of indicators developed by the national authorities responsible for garden cities, environmental protection role model cities, clean cities, eco-cities, and eco-friendly garden cities, serving as a reference that was discussed and analyzed before grouped into Levels 1, 2, and 3 indicators based on assessment metric principles. These efforts led to the assessment metrics for environmentally friendly cities below (Table 3.1). These metrics are divided into three levels. Level 1 aims to comprehensively assess environmentally friendly cities to determine their overall environmental friendliness, the primary objective. Level 2 is composed of indicators for economic development, environmental protection, and social progress, which need to be evaluated and controlled to achieve the overall objective. Level 3 includes specific indicators for assessing the status of specific targets.

[8] Li et al. [7].

3.2.2.1 Specific Indicators Include

The Economic Development Pattern The percentage of GDP in the tertiary sector: the percentage of the regional GDP in the tertiary sector. This indicator measures advancement of the regional industry mix.

The percentage of high-tech industries out of the total industrial output value: the percentage of the region's industrial output value that is derived from high-tech industries. This indicator reflects the size of high-tech industries and the region's economic structure.

The percentage of industrial water reused: the percentage of total industrial water that is reused. The industrial sector consumes massive amounts of water and has a centralized water supply, providing greater potential for water conservation.

The amount of chemical fertilizers used per unit area of farmland: the amount of chemical fertilizers used for agriculture per unit area of farmland across the region in the given year. This indicator reflects potential environmental pressures from agricultural production. Chemical fertilizer quantities are converted into the amount of active ingredients, such as nitrogen, phosphorus pentoxide, and potassium oxide, calculated according to their respective percentages in the fertilizer. The equation is:

$$\text{The amount of chemical fertilizers used per unit area of farmland} = \\ \text{the amount of chemical fertilizers used}\left(\text{kg}\right) / \text{farmland area}\left(\text{ha}\right)$$

Energy consumption per unit of GDP: energy consumption per 1,000 yuan of GDP in the region. This indicator reflects the region's energy efficiency and economic sustainability.

Water consumption per unit of GDP: the ratio of the city's total water consumption to its GDP. Total water consumption in the city includes industrial and domestic water consumption and excludes agricultural and ecosystem water consumption. Industrial water consumption only refers to freshwater consumption. Reducing water consumption per unit of GDP key to sustainable water use in China. The equation is: water consumption per unit of GDP (m^3/1,000 yuan) = total water consumption in the city (m^3)/its GDP (1,000 yuan).

Land used per unit of GDP reflects ecological constraints unique to the city given the scarcity of urban land resources, one of the major bottlenecks in socioeconomic development. An uncontrolled increase in the area of land used for industrial purposes accompanied by scattered construction and low output efficiency often plagues and hinders urban economic development.[9]

The percentage of industrial solid waste treated and reused: the percentage of industrial solid waste treated and reused out of the total amount of industrial solid waste generated. Dumping industrial solid waste uses a lot of land and causes secondary air, surface water, and groundwater pollution. Reducing industrial solid waste is an important part of China's efforts to control the total amount of pollutants

[9] Zhou et al. [8].

Table 3.1 Assessment metrics for environmentally friendly cities

Level 1 indicators	Level 2 indicators	No.	Level 3 indicators
Comprehensive assessment of environmentally friendly cities	I. Economic development	1	The percentage of GDP in the tertiary sector (%)
		2	The percentage of high-tech industries out of total industrial output value (%)
		3	The percentage of reused industrial water (%)
		4	The amount of chemical fertilizers used per unit area of farmland (converted into the amount of active ingredients) (kg/ha)
		5	Energy consumption per unit of GDP (tons of standard coal/thousand yuan)
		6	Water consumption per unit of GDP (m^3/thousand yuan)
		7	Land use per unit of GDP (ha/billion yuan)
		8	The percentage of industrial solid wastes treated and reused (%)
		9	Forest coverage (%)
		10	The average area of public green spaces per capita (m^2/person)
		11	The average amount of freshwater resources per capita (m^3/person)
	II. Environmental protection	12	The percentage of nature reserves out of total land area (%)
		13	The average area of farmland per capita (ha/person)
		14	The percentage of good air quality days in a year (%)
		15	The percentage of compliant water quality in functional zones of the urban aquatic environment (%)
		16	The percentage of compliant water quality in centralized drinking water sources (%)
		17	The percentage of urban domestic sewage centrally treated (%)
		18	The percentage of compliant industrial wastewater discharged (%)
		19	The average ambient noise level in urban areas (db)
		20	The SO_2 emission intensity (kg/thousand yuan) (GDP)
		21	The percentage of MSW treated into harmless substances (%)
		22	The percentage of hazardous waste that is treated (%)
		23	The percentage of environmental investment in GDP (%)

		24	The percentage of motor vehicles regularly checked for environmental purposes (%)
		25	The percentage of clean energy sources (%)
		26	The percentage of projects that have undergone EIA (%)
		27	The percentage of motor vehicles with compliant exhaust emissions (%)
	III. Social progress	28	The Engel coefficient (%)
		29	Public satisfaction with urban environmental protection (%)
		30	The penetration of environmental education in elementary and middle schools (%)

generated. Reusing industrial solid waste involves extracting usable resources, energy, and other raw materials from waste or converting it into resources, energy, and raw materials through recovery, processing, circulation, exchange, or other methods. Industrial solid waste may be treated and reused as agricultural fertilizers and as materials for production, construction, and roads.

Environmental Protection Forest coverage: the percentage of total land area covered by forests, representing the regional forest resources and woodlands. Forests covering 30 % of total land area in an even distribution are said to be necessary for effective regulation of the local climate and maintaining ecological agricultural balance in a country or region. A country or region with such extensive coverage is said to have a good natural environment.

The average area of public green spaces per capita: the average area of public green spaces per urban resident. This important indicator reflects urban development, landscaping, and the environment.

The average amount of freshwater resources per capita: the ratio of total surface water and groundwater controlled by humans and directly available in the region to total population. This indicator reflects dynamic changes in the amount of regional water resources.

The percentage of nature reserves out of the total land area: the percentage of the total area of nature reserves, scenic spots, and forest parks out of the total urban land area (including districts and district-level cities administered by the municipality). This indicator reflects regional efforts to protect species, genes, and ecosystems as well as regional resource sustainability.

The average farmland area per capita: the ratio of the total area of land planted with crops to the region's total population. This important indicator measures the agricultural sustainability and reflects regional efforts in dealing with arable land resource shortages, the effectiveness of such efforts, and resource autonomy. China's current average farmland area per capita is only 0.093 acres, 40 % of the world average, presenting a grim reality for farmland in China. Further urbanization deteriorates the situation.

The percentage of good air quality days in a year: the percentage of days with air quality Level 2 or higher (the air quality index (AQI) at Level 2 or higher) in a year.

The percentage of compliant water quality in functional zones of the urban aquatic environment: the weighted average of the percentages of compliant water quality in different functional zones, calculated by monitoring results from specific bodies of surface water in urban areas and from specified points along the coast for coastal cities. This indicator is critical for assessing urban surface water quality. The equation is:

$$\text{The percentage of compliant water quality at given sections} = \text{the sum compliant measurements at specified sections} / \text{the total number of measurements at specified sections} \times 100\ \%$$

The percentage of compliant water quality in centralized drinking water sources: the percentage of water from urban centralized drinking water sources that meets Level 3 quality requirements from the *Surface Water Quality Standard GB3838-2002* or the *Groundwater Quality Standards GB/T14848-1993* dependent on the water source.

The percentage of urban domestic sewage out of total wastewater discharged that is centrally treated: the percentage of sewage that receives Level 2 treatment or better at urban sewage treatment plants, meeting the national discharge standard.

The percentage of compliant industrial wastewater discharged: the percentage of industrial wastewater out of total wastewater discharged that is in accordance with the national discharge standard after treatment. This indicator reflects regional efforts in industrial pollution prevention and the impact on surrounding bodies of water.

The average ambient noise level in urban areas: the average sound levels monitored in a certified ambient noise grid in a built-up area of the city. This indicator helps control noise pollution and reduces the ambient noise level in the city, creating a quiet working and living environment for the general public.

The SO_2 emission intensity: the amount of SO_2 generated per unit of GDP in a year. The general requirements for energy conservation and emissions reduction offer the following equation to calculate this indicator:

$$\text{The } SO_2 \text{emission intensity} = \text{the total amount of } SO_2 \text{emitted in the year}(\text{kg}) / \text{the city's annual GDP}(\text{thousand yuan})$$

SO_2 emissions must remain below the overall target specified by the state and must have decreased continuously over the past 3 years

The percentage of MSW treated into harmless substances: the percentage of MSW treated into harmless substances out of the total amount of MSW generated in urban areas of the city.

The percentage of hazardous waste treated: the percentage of hazardous waste (including hazardous industrial waste, medical waste, etc.) treated out of the total amount generated. This indicator can be used to assess the percentage of medical waste treated in a centralized manner (or industrial hazardous waste treated and reused) out of the total amount generated in areas administered by the city in the year. The equation is:

$$\text{The percentage of hazardous industrial waste treated and reused} = \text{the amount of hazardous industrial waste treated and reused}(\text{tons}) / \text{the total amount of hazardous industrial waste generated}(\text{tons}) \times 100\ \%$$

The amount of hazardous industrial waste treated and reused includes the amounts treated and reused within the city and of the waste legally moved elsewhere for treatment and reuse in the year.

Any and all used radiation sources and radioactive waste must be safely disposed of in accordance with law. Any and all users of radioisotopes and ray devices must hold a radiation safety license.

The percentage of environmental investment in GDP: the percentage of total investment in environmental pollution prevention and control and environmental protection and development in the annual GDP. In China this percentage is reported to have remained unchanged or to have continuously increased over the past 3 years.

The percentage of motor vehicles regularly checked for environmental purposes: the percentage of motor vehicles that have been checked for environmental compliance out of the total number of motor vehicles registered in the city in the reference year and the number of motor vehicles, including automobiles and motorcycles, that have received environmental checks from an inspection agency designated by the provincial department of environmental protection using methods in accordance with the *Air Pollution Prevention Law* and relevant MEP documents.

The percentage of clean energy used: the percentage of clean energy sources used (measured in tons of standard coal equivalent) out of the total energy consumption in urban areas of the city. Total energy consumption covers the various energy sources used in production and consumption activities across the city minus the energy used and lost during processing and conversion activities intended to create secondary energy sources over a certain period of time. Clean energy sources in the city include natural gas, coke oven gas, other coal gas, flare gases from refineries, liquefied petroleum gas, and other clean gases, as well as diesel power generation, low-sulfur light diesel oil, and other clean fuels (excluding motor vehicle fuels) used for combustion.

The percentage of projects that have undergone EIA: the percentage of projects that have undergone EIA out of the total number of projects that were launched in the city over the past 3 years.

The percentage of motor vehicles with compliant exhaust emissions: the percentage of automobiles and motorcycles that have passed the annual exhaust inspection out of the total number of automobiles and motorcycles registered at the municipal transport authorities in urban areas of the city.

Social Progress The Engel coefficient: the percentage of food costs out of total consumer spending. The lower the coefficient, the higher the standard of living.

Public satisfaction with urban environmental protection: public satisfaction with environmental protection efforts and environmental quality.

The penetration of environmental education in elementary and middle schools: the percentage of elementary and middle schools that have environmental education out of the total number of elementary and middle schools in the city.

Determining Indicator Reference Values After the assessment metrics for environmentally friendly cities are established, the challenge of grading and quantifying reference values for each indicator arises. The grading criteria should reflect the city's development toward environmentally friendly standards as accurately as possible. They should also be able to reflect interregional differences and adapt to local conditions. Specific reference values should be established in accordance with the following principles: (1) The reference values of existing national or international standards should be adopted whenever possible. The reference value of each indicator should rely primarily on relevant national environmental assessment criteria, such as the *National Standard for Assessing Cities as Role Models in Environmental Protection and Their Implementation Rules*; the *National Standard for Eco-cities and Garden Cities*; the *National Standard for Sanitary Cities*; the *Indicators of the Development of Eco-districts, Eco-cities, and Eco-provinces*; and the *Implementation Rules for Assessment Indicators of Comprehensive Urban Environmental Improvement in the 11th Five-Year Plan Period.* Other references include criteria found in other works and literature.[10,11,12,13,14,15] (2) Current environmental status of global environmentally friendly cities should be used as reference. Existing reference values should be quantified whenever possible and be consistent with the current levels of environmental and socioeconomic development. (3) Indicators for which complete statistics are unavailable but which are essential for assessment may be temporarily determined by consulting experts. Existing reference values do not necessarily represent ultimate targets of environmentally friendly urban development, as they can only reflect what is attainable under current socioeconomic and technological conditions in Chinese cities. The reference values for indicators of current environmentally friendly urban development were established using these principles (see Table 3.2).

3.3 Measures for Building Environmentally Friendly Cities

Environmentally friendly cities are complex, long-term, and systematic urban development projects, whose implementation should first target economic, social, and environmental systems, focusing on transforming the economic growth pattern by developing a circular economy. Initial efforts should concentrate on basic urban development elements, such as science, technology, culture, and eco-civilization.

[10] Zhang and Zheng [5].

[11] Zhang [4].

[12] Wen and Li [9].

[13] Huang [10].

[14] Huang [10].

[15] Wu et al. [11].

Table 3.2 Grading criteria for assessment indicators of environmentally friendly cities

No.	Level 3 indicator	Low	Medium	High	Indicator type
1	The percentage GDP in the tertiary sector (%)	≥35	≥40	≥60	Positive
2	The percentage of high-tech industries out of the total industrial output value (%)	≥25	≥30	≥50	Positive
3	The percentage reused industrial water (%)	≥40	≥60	≥80	Positive
4	The amount of chemical fertilizers used per unit area of farmland (converted into the amount of active ingredients) (kg/ha)	≤380	≤350	<250	Negative
5	Energy consumption per unit of GDP (tons of standard coal/thousand yuan)	≤1.2	≤0.9	≤0.84	Negative
6	Water consumption per unit of GDP (m^3/thousand yuan)	150	90	60	Negative
7	Land used per unit of GDP (ha/billion yuan)	20	TBD	TBD	Negative
8	The percentage of industrial solid waste treated and reused (%)	≥70, with no discharge of hazardous waste	≥85, with no discharge of hazardous waste	≥90, with no discharge of hazardous waste	Positive
9	Forest coverage (%)	≥65 in mountainous areas; ≥35 in hilly areas; ≥10 in plains; forest/grass coverage ≥80 in frigid or grassland zones	≥70 in mountainous areas; ≥40 in hilly areas; ≥15 in plains; forest/grass coverage ≥85 in frigid or grassland zones	≥75 in mountainous areas; ≥45 in hilly areas; ≥20 in plains; forest/grass coverage ≥90 in frigid or grassland zones	Positive
10	The average area of public green spaces per capita (m^2/person)	≥8.5	≥12	≥16	Positive

(continued)

Table 3.2 (continued)

No.	Level 3 indicator	Low	Medium	High	Indicator type
11	The average amount of freshwater resources per capita (m^3/person)	≥1,000	≥2,000	≥3,000	Positive
12	The percentage of nature reserves out of the total land area (%)	5	12	≥17	Positive
13	The average area of farmland per capita (ha/person)	≥0.060	≥0.075	≥0.110	Positive
14	The percentage of good air quality days in a year (%)	≥85	≥90	≥95	Positive
15	The percentage of compliant water quality in functional zones of the urban aquatic environment (%)	100	100	100	Positive
16	The percentage of compliant water quality in centralized drinking water sources (%)	100	100	100	Positive
17	The percentage of urban domestic sewage centrally treated (%)	≥80	≥85	≥90	Positive
18	The percentage of compliant industrial wastewater discharged (%)	100	100	100	Positive
19	The average ambient noise level in urban areas (db)	≤60	≤56	<50.0	Negative
20	The SO_2 emission intensity (kg/thousand yuan) (GDP)	<6	≤5 and within the overall target prescribed by the state	≤4.5	Negative
21	The percentage of MSW treated into harmless substances (%)	≥70	≥85	≥90	Positive
22	The percentage of hazardous waste treated (%)	100	100	100	Positive

(continued)

Table 3.2 (continued)

No.	Level 3 indicator	Low	Medium	High	Indicator type
23	The percentage of environmental investment in GDP (%)	≥1.2	≥2.4	≥3.5	Positive
24	The percentage of motor vehicles regularly checked for environmental purposes (%)	≥80	≥85	≥95	Positive
25	The percentage of clean energy sources (%)	≥50	≥60	≥70	Positive
26	The percentage of projects that have undergone EIA (%)	≥85	≥90	≥100	Positive
27	The percentage of motor vehicles with compliant exhaust emissions (%)	≥70	≥80	≥95	Positive
28	The Engel coefficient (%)	≤40	<30	≤20	Negative
29	Public satisfaction with urban environmental protection (%)	>60	≥85	>90	Positive
30	The penetration of environmental education in elementary and middle schools (%)	≥80	≥85	>90	Positive

3.3.1 Transforming the Economic Growth Pattern by Developing a Circular Economy

Transforming the economic growth pattern by developing a circular economy rests at the core of efforts toward an environmentally friendly city. A fundamental difference between a circular economy and a traditional economy is its rejection of a one-directional linear development flow that leads from resources to products to pollution and favors instead a feedback loop of resources turned products and renewed as resources. All substances and energies of this ongoing circulation are used in a reasonable and sustainable manner to minimize the impact of economic activities on the natural environment.[16,17] Wealth is increased at the expense of natu-

[16] Ma [12].

[17] Xu [13].

ral resources and the environment under the traditional economic development pattern, so that it accompanies high consumption and pollution and produces few benefits. The fundamental drawback of this pattern is its inability to coordinate economic, resource, and environmental system development leading to economic development based on the overexploitation of natural resources and environmental pollution and damages. The inevitable consequence is unsustainable economic development due to the diminished environmental support and growing conflicts between economic activities and the environment. The economic development-driven increased demand for natural resources places a price on them as scarce economic resources. A shrinking supply of overexploited and polluted natural resources is further aggravating the scarcity of these resources.[18] The traditional economic development pattern will inevitably lead to intense environmental deterioration, an increasing shortage of resources and energy, and a continuous decline in the people's happiness index. China has seen rapid economic growth since it began reforming and opening up although it has had to contend with a large population and a relative shortage of resources. China is facing severe obstacles to further socioeconomic development under the traditional development pattern, such as basic resource depletion and continued environmental deterioration. China's energy, water, and land consumption per unit of GDP is much higher than that of developed countries and the world average. China requires much progress in environmental governance due to complex structural environmental problems. It will be difficult for China to achieve sustainable socioeconomic development if it continues with the traditional economic development pattern and environmental governance focused on treating emissions and discharge.[19] On October 16, 2002, the then Chinese President Jiang Zemin spoke at the Second Global Environment Facility (GEF) Assembly, stating that sustainable development could be achieved only by developing a circular economy based on the most efficient use of resources and environmental protection. In 2003, the then Chinese President Hu Jintao spoke at the central government's meeting on population, resources, and the environment, stressing that China should accelerate economic transformation; promote a circular economy throughout regional economic development, urban and rural development, and production activities to most efficiently use resources; minimize the quantity of waste generated and gradually restore the virtuous cycle of ecosystems; and create cities that serve as role models in environmental protection, eco-civilization demonstration zones, and eco-provinces. The *Decision on Strengthening the Governing Capacity of the Communist Party of China* passed at the 4th plenary session of the 16th CPC Central Committee officially established resource conservation, environmental protection, the development of a circular economy, and the formation of a conservation-oriented society as important tasks in implementing the Scientific Outlook on Development and enhancing its ability to manage the socialist market economy. In 2005, the then Chinese Premier Wen Jiabao spoke at the government work report at the 3rd session of the 10th NPC, stressing that "we will energetically

---

[18] Wu and Dong [14].

[19] Wang [15].

develop the recycling sector of the economy. We will increase comprehensive resource utilization and recycling by addressing problems concerning resource exploitation, production use, and social use, and the utilization of waste materials. We will also work vigorously to develop new and renewable energy resources." A major reason for the rapid popularization the circular economy in China is it is an objective need for comprehensively building a moderately prosperous society.[20] Replacing the existing economic development pattern with an environmentally friendly pattern is the primary task for building environmentally friendly cities.

3.3.2 Developing Green Technology

Green technology is the technical support for an environmentally friendly society. Technological history is full of concepts and ideas intended to fight and conquer nature. Existing technologies have given humans the capability to rob nature of resources and exacerbated exploitation, leading nature to retaliate. Traditional industrial technologies have caused resource scarcity, pollution, and unsustainable development; however, green technology can enable resource abundance, purity, and sustainability. Traditional technological ideas favoring confrontation and conquest must be overcome to make way for symbiosis and coexistence of humanity and nature, in order to create environmentally friendly cities.[21],[22] Policymakers and city leaders should work toward the best guidance and control when constructing an environmentally friendly city. Relevant organizations should enhance technical support by developing technologies that enable revenue increase, cost reduction, and environmental protection. Green technology innovation capacity increase and application of proven green technologies are also crucial in this process. Relevant organizations should also promote technological progress in improving the entire product lifecycle, creating a production system with low resource and energy consumption, high efficiency, and minimal waste generation.[23]

3.3.3 Developing Environmental Culture and Eco-civilization

Environmental culture and eco-civilization underpin an environmentally friendly city. They both seek harmonious relationships between humanity and nature and among humans, as well as sustainable development.[24] Traditional industrial civilization must be transformed in an eco-civilization, including widespread environmen-

[20] Xie [16].

[21] Pan [17].

[22] Ren et al. [18].

[23] Zhang [4].

[24] Li and Li [19].

tally friendly values and morality throughout the economy, technology, law, ethics, and politics, which pursues harmony between humanity and nature and among people if an environmentally friendly city is to be built.[25] Forming an environmental culture requires strengthening research, development, and promotion of green technology to provide technical support for environmental protection and public awareness and participation. An organic combination of environmental cultural development and environmental ethics education and increased environmental awareness in administration, economy, law, science, etc. are also crucial to instilling environmental culture in various fields.

3.3.4 Promoting Green Politics

Environmental issues involve a variety of social issues that cover multiple fields such as the environment, resources, population, the economy, technology, and culture. Activities across society must be guided toward sustainable development and governmental social management service functions changed in order to solve environmental problems. Environmental issues are hence inherently political.[26] Effective solutions to environmental issues are needed at the political level, and ecology and political science must combine to promote green politics. Governments lead and organize the city and have a responsibility to advance green politics and facilitate sustainable urban development, harmony, and symbiosis between humanity and nature.[27]

3.3.5 Building a Healthy Environment

Building an environmentally friendly city grants people access to clean water and air, safe foods, and a healthy environment for work and life. A healthy natural environment with preserved ecosystems is the foundation for human survival and development and is an ever-present goal for humanity. Environmental deterioration induces human diseases and directly threatens human health. Environmental deterioration primarily results from ecological crises caused by conflicts between humans and nature. Only a deep understanding of relations between humans and nature and improved relations that maintain harmony can guarantee human health.[28] Respecting the environment or ecosystem's carrying capacity and natural laws and adopting production methods, lifestyles, and consumption patterns that promote a resilient environment are necessary for the creation of an environmentally friendly

[25] Yu and Ren [20].

[26] Pan [21].

[27] Zhao [22].

[28] Wei [23].

city. Human activities' adverse impact on the environment should be minimized to sustain the supply of natural resources and the ecosystems' waste absorbing and recycling capacities. This promotes a virtuous circle in which human activities maintain environmental health, which then promotes humanity's socioeconomic development.

3.3.6 Advocating for Environmentally Friendly Consumption Patterns

Green, just, moderate, and resources-conserving consumption patterns sustainably favor eco-civilization and are critical for developing environmentally friendly society. Traditional consumption patterns convert resources to products to waste in a one-way process. These patterns exhaust natural resources by converting them into industrial and agricultural products or consuming them directly to meet people's growing material needs. Environmentally friendly consumption is a sustainable pattern where humans constrain consumption within the environment's carrying capacity, providing conducive circumstances for environmental protection and ecological balance without damaging future generations' interests. Environmentally friendly consumption requires optimal and sustainable resource use and minimal waste generation and environmental pollution. Just and moderate consumption implies that each individual, whether of current or future generations, is equally entitled to a good quality of life. This consumption pattern first addresses people's needs and health and seeks intragenerational and intergenerational justice. A simple life with reduced consumption that has a negligible on human health is important in this process. Green consumption ensures that the needs, safety, and health of current and future generations will be satisfied. China's increasing resource and environmental constraints compel the Chinese government to increase public awareness of the ecological crisis and of green, low-carbon, or environmentally friendly consumption. It should encourage conservation throughout all actions down to the individual level so that every person will conserve energy, water, materials, and food and recycle garbage.[29] These strategies all attempt to create harmony between humanity and nature by developing a circular economy, green technology, environmental culture, an eco-civilization, green politics, and a healthy environment while promoting environmentally friendly consumption.

An environmentally friendly economic development pattern, green political systems, environmentally friendly cultural values, green technology, and a healthy natural environment are basic elements in creating environmentally friendly cities and societies. Fundamental societal elements include population, the natural environment, the economy, politics, ideology, and culture. The social organism's existence and development are inseparable from these elements. Society as a whole and

[29] Li and Li [19].

its separate units are produced as these factors combine.[30] Environmentally friendly cities are the starting point and foundation of an environmentally friendly society. Environmentally friendly cities are the focus of a new urban development plan and are necessary for a society that conserves resources and is environmentally friendly, as well as for implementing the Scientific Outlook on Development and for coordinating economic, social, and environmental development in China.

3.4 Shanghai, An Environmentally Friendly City

Shanghai is the economic, financial, trading, and shipping center in Mainland China, leading to predictable conflicts between further economic development and resource and environmental sustainability. The fuzzy aggregation index method (FAIM) was used to analyze Shanghai's environmentally friendly urban development. The FAIM is a fuzzy inference analysis and evaluation method that covers qualitative and quantitative data and accounts for varying levels of precision. It describes transitional states in social and economic phenomena using fuzzy terms, concepts with a clear intension but unclear extension, and offers comprehensive assessment and evaluation of multiple factors. Environmental friendliness is a fuzzy concept from a mathematical perspective, having a clear intension but unclear extension, so that a city's level of environmental friendliness is best described using fuzzy subset and membership. The system for evaluating an environmentally friendly city is complex and applicable on multiple levels, which warrants using the FAIM to reflect the systematic organization and various levels of environmentally friendly social assessment. A multilevel fuzzy comprehensive evaluation model is often used in the comprehensive evaluation process. This model can solve at least two issues. First, it enables comprehensively sorting cities into different regions with respect to the level of environmental friendliness, so that the evaluation results are comparable between regions. Second, it enables evaluation of cities in a particular region with respect to the level of environmental friendliness.[31] The comprehensive evaluation index (CEI) of Shanghai for 2005–2010 with respect to economic development, environmental protection, social progress, and environmental friendliness was calculated using the Level 3 indicators from Table 3.2 (see Table 3.3 and Fig. 3.1).

Table 3.3 and Fig. 3.1 show how Shanghai's levels of economic development, environmental protection, social progress, and CEI were all greater than 1 in 2005–2010 according to the low-level criteria for environmentally friendly urban development. Most value follows an upward trend from year to year. The values and trends suggest that Shanghai entered the initial stage of environmentally friendly urban development during this period. Economic development still maintains a greater value than environmental protection and social progress. The city's economic development received more attention than the other two aspects during this period, unbalancing integrated

[30] Wu et al. [24].

[31] Hua and Pan [25].

Table 3.3 Values of Level 1 and 2 indicators for Shanghai corresponding to Level 3 criteria for environmentally friendly urban development

Level 3 criteria	Levels 1 and 2 indicators	2005	2006	2007	2008	2009	2010
Low	Economic development	1.21	1.23	1.32	1.40	1.47	1.67
	Environmental protection	1.11	1.15	1.18	1.21	1.28	1.25
	Social progress	1.11	1.12	1.13	1.09	1.14	1.15*
	CEI	1.14	1.17	1.21	1.23	1.30	1.36
Medium	Economic development	0.86	0.87	0.93	0.98	1.02	1.16
	Environmental protection	0.83	0.87	0.89	0.94	0.98	1.01
	Social progress	0.84	0.84	0.85	0.82	0.86	0.88*
	CEI	0.84	0.86	0.89	0.91	0.95	1.02
High	Economic development	0.65	0.62	0.66	0.69	0.73	0.81
	Environmental protection	0.73	0.76	0.78	0.83	0.87	0.90
	Social progress	0.56	0.56	0.56	0.55	0.57	0.58*
	CEI	0.64	0.65	0.67	0.69	0.72	0.76

Note: *indicates that the number is an interpolation, not real, due to lack of complete data. A number greater than 1 suggests that this city is already in the given stage. Otherwise, this city has not entered the given stage. The smaller the number, the farther it is from the given stage

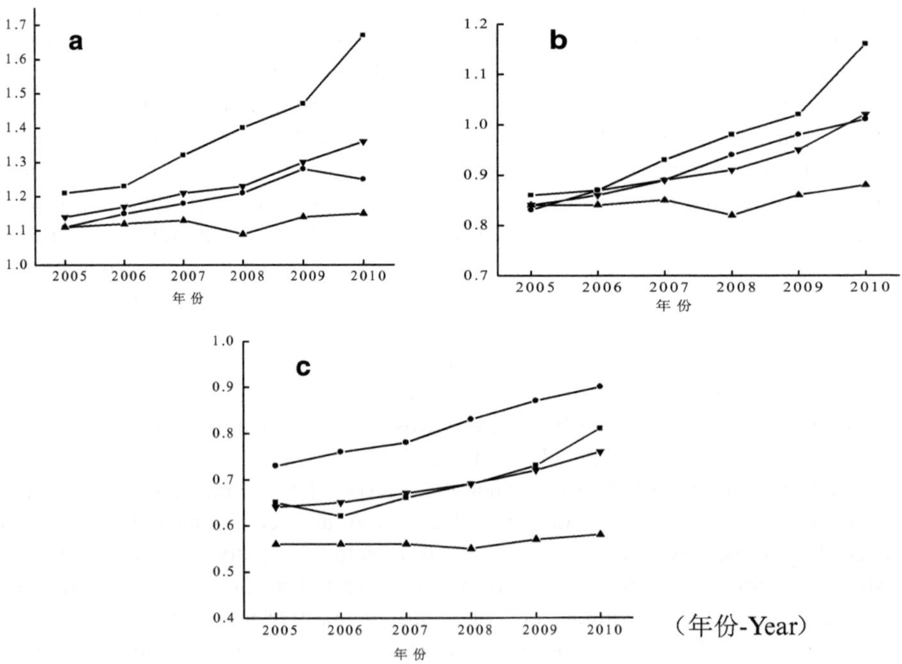

Fig. 3.1 Changes in Shanghai's levels of economic development, environmental protection, social progress, and CEI over time (**a**, **b**, and **c** correspond to low, medium, and high, respectively; economic development, environmental protection, social progress, and CEI are denoted by ■, ●, ▲, and ▼, respectively)

Table 3.4 Top 50 Chinese cities by CEI of environmentally friendly city (2008–2011)

2008		2009		2010		2011	
City	Ranking	City	Ranking	City	Ranking	City	Ranking
Shenzhen	1	Shenzhen	1	Guangzhou	1	Shenzhen	1
Nanjing	2	Guangzhou	2	Shenzhen	2	Guangzhou	2
Beijing	3	Beijing	3	Shanghai	3	Shanghai	3
Shanghai	4	Shanghai	4	Beijing	4	Beijing	4
Guangzhou	5	Nanjing	5	Nanjing	5	Nanjing	5
Zhuhai	6	Hangzhou	6	Hangzhou	6	Hangzhou	6
Nanchang	7	Xiamen	7	Tianjin	7	Zhuhai	7
Hangzhou	8	Tianjin	8	Zhuhai	8	Xiamen	8
Tianjin	9	Zhuhai	9	Xiamen	9	Tianjin	9
Taiyuan	10	Wuhan	10	Suzhou	10	Ningbo	10
Xiamen	11	Suzhou	11	Dalian	11	Wuxi	11
Qingdao	12	Nanchang	12	Nanchang	12	Suzhou	12
Chengdu	13	Hefei	13	Jinan	13	Xi'an	13
Yinchuan	14	Dalian	14	Changsha	14	Wuhan	14
Hefei	15	Dongying	15	Dongying	15	Changzhou	15
Jinan	16	Changsha	16	Wuhan	16	Dalian	16
Changsha	17	Nanning	17	Hefei	17	Jinan	17
Suzhou	18	Wuxi	18	Changchun	18	Changsha	18
Wuxi	19	Jinan	19	Shenyang	19	Daqing	19
Nanning	20	Changchun	20	Hohhot	20	Chengdu	20
Changchun	21	Hohhot	21	Yinchuan	21	Shenyang	21

(continued)

Table 3.4 (continued)

2008		2009		2010		2011	
City	Ranking	City	Ranking	City	Ranking	City	Ranking
Zhengzhou	22	Taiyuan	22	Yantai	22	Qingdao	22
Dalian	23	Fuzhou	23	Qingdao	23	Dongying	23
Zhongshan	24	Chengdu	24	Weihai	24	Weihai	24
Hohhot	25	Qingdao	25	Xi'an	25	Yantai	25
Changzhou	26	Weihai	26	Wuxi	26	Changchun	26
Weihai	27	Changzhou	27	Fuzhou	27	Nanchang	27
Shenyang	28	Yantai	28	Chengdu	28	Hefei	28
Dongying	29	Zhengzhou	29	Nanning	29	Yinchuan	29
Yantai	30	Ningbo	30	Quanzhou	30	Fuzhou	30
Fuzhou	31	Yinchuan	31	Taiyuan	31	Zhengzhou	31
Chongqing	32	Zhongshan	32	Changzhou	32	Chongqing	32
Wuhan	33	Chongqing	33	Baoding	33	Taiyuan	33
Xi'an	34	Quanzhou	34	Ningbo	34	Hohhot	34
Ningbo	35	Shenyang	35	Zhengzhou	35	Shaoxing	35
Lanzhou	36	Shaoxing	36	Daqing	36	Kunming	36
Qinhuangdao	37	Xi'an	37	Chongqing	37	Harbin	37
Ma'anshan	38	Qinhuangdao	38	Shaoxing	38	Nanning	38

Tai'an	39	Linyi	39	Harbin	39	Linyi	39
Lianyungang	40	Tai'an	40	Zhongshan	40	Huzhou	40
Huzhou	41	Baoding	41	Shantou	41	Qinhuangdao	41
Shaoxing	42	Daqing	42	Lanzhou	42	Shijiazhuang	42
Xiangtan	43	Harbin	43	Tai'an	43	Guiyang	43
Shantou	44	Lanzhou	44	Guiyang	44	Lanzhou	44
Guilin	45	Shijiazhuang	45	Huzhou	45	Quanzhou	45
Linyi	46	Shantou	46	Shijiazhuang	46	Tai'an	46
Baoding	47	Huzhou	47	Kunming	47	Zhongshan	47
Shijiazhuang	48	Guiyang	48	Qinhuangdao	48	Handan	48
Guiyang	49	Kunming	49	Handan	49	Shizuishan	49
Baoji	50	Guilin	50	Lianyungang	50	Ma'anshan	50

and coordinated urban development. Shanghai's levels of economic development, environmental protection, social progress, and CEI were rising during 2005–2010 according to the medium-level criteria as well. Economic development and environmental protection levels were greater than 1 in 2009 and 2010. The CEI remained below 1 during the entire period, suggesting that Shanghai had not yet entered the intermediate stage of environmentally friendly urban development. Shanghai's levels of economic development, environmental protection, social progress, and CEI rose during 2005–2010 based on the high-level criteria, but were all significantly smaller than 1, suggesting that Shanghai has much work ahead before entering the advanced stage of environmentally friendly urban development (Table 3.4).

References

1. Ren Y, Yu H, Xia G et al (2005) The concept of environmentally friendly society: knowledge base and meaning. Environ Econ (12):17–25
2. Song Y (2008) On an environmentally friendly society from a philosophical perspective. J Tianzhong (6):31–33
3. Fang S (2006) On building an environmentally friendly society from a philosophical perspective. Philos Res (4):114–116
4. Zhang Y (2008) Preliminary research on theories and assessment methodologies regarding an environmentally friendly society. A thesis for a doctoral degree from East China Normal University, Shanghai
5. Zhang X, Zheng Z (2007) On the environmental metrics for environmentally friendly cities. Environ Sci Manag (9):53–56
6. Su M, Yi W (2003) On building a sustainable eco-city. Safety Environ Eng (2):12–15
7. Li J, Li M, Wang J et al (2010) Greenbook of environment competitiveness: report on China's provincial environment competitiveness development (2005–2009), Social Sciences Academic Press, Beijing
8. Zhou F, Jing K, Liu X (2008) Shanghai sustainable development report 2008: urban ecological civilization. Xuelin Press, Shanghai
9. Wen Z, Li L (2007) On environmentally friendly cities: metrics and benchmarking. Environ Prot (8):26–28
10. Huang X (2010) On the relative environmental friendliness index: measurement and integrated control mechanism. A thesis for a doctoral degree from China University of Technology, Hefei
11. Wu M, Zhang Y, Lin F (2010) .Assessment on environmentally friendly society from a social structure perspective. Resour Sci (11):2100–2105
12. Ma C (2006) On developing a circular economy and building a resource-saving, environmentally friendly society. Shandong Soc Sci (2):15–18
13. Xu T (2007) Environmentally friendly society: meaning and recommendations on measures. J Qingdao Univ of Sci and Technol (Soci Sci) (1):82–86
14. Wu Y, Dong S (2000) On institutional innovation for implementing the sustainable development strategy: building a green economic system, World Environ (4):14–17
15. Wang Y (2007) Achieve sustainable socioeconomic development by building a resource-saving, environmentally friendly society. Bull Chin Acad Sci (2):147–152
16. Xie Z Developing a circular economy is an important approach to building environmentally friendly society. People's Daily online, http://finance.people.com.cn/GB/1037/3751145.html
17. Pan Y, On how environmental protection relates to technology. the Ministry of Environmental Protection, http://www.mep.gov.cn/gkml/hbb/qt/200910/t20091030_180604.htm

18. Ren Y, Yu H, Xia G et al (2006) The concept of an environmentally friendly society: knowledge base, meaning and approach. A collection of excellent papers from the annual academic conference of Chinese Society for Environmental Sciences, Beijing, pp 49–54
19. Li X, Li W (2006) Environmentally friendly cities. China Environmental Science Press, Beijing
20. Yu H, Ren Y (2006) Environmentally friendly society: historical logic, core foundation and approach. Environ Prot (8):56–59
21. Pan Y (2006) Harmonious society vs. environmentally friendly society. Environ Prot (7):4–17
22. Zhao C (2007) The need for green politics: a sustainable world with human-nature harmony and symbiosis. A thesis for a doctoral degree from Jilin University, Changchun
23. Wei M (2006) On how the environment relates to human health. New Heights (4):94–96
24. Wu Z, Wu P, Su Z (2005) Modern sociology. Shanghai People's Publishing House, Shanghai
25. Hua H, Pan Y (2005) A review of sustainable development assessment methods. J Yunnan Normal Univ (3):65–70

Chapter 4
Resource-Conserving Cities

Lingfeng Kang and Youyi Zhao

4.1 Resource-Conserving City: Concept and Definition

4.1.1 Concept

A resource-conserving city takes comprehensive measures to improve resource utilization in production, distribution, and consumption to maximize economic and social benefits using minimal resources and to achieve sustainable socioeconomic development.[1] This new urban development concept emphasizes the needs of current and future urban development through reduced resource consumption and improved resource utilization and by ensuring urban economic efficiency and people's quality of life. A resource-conserving city requires sustainable socioeconomic development and harmony between humanity and nature achieved by prioritizing resource conservation over resource exploitation and by shifting the traditional economic growth and consumption patterns to optimize resource and energy allocation. A resource-conserving city seeks maximum production output with minimum input, improved production efficiency and product quality, and reduced natural resource consumption per unit of output. It reduces dependency on resources and obtains greater benefits whenever possible and as long as certain consumption requirements are met. A resource-conserving city establishes a virtuous circle between production and consumption and a demand-oriented production mechanism. It aims to lower resource consumption, reduce environmental pollution, and accumulate socioeconomic benefits. All production, consumption, and distribution activities

[1] Kanjing [1].

L. Kang (✉) • Y. Zhao
Lanzhou City University, Lanzhou, China
e-mail: 1018570653@qq.com; 1369802664@qq.com

J. Li, T. Yang (eds.), *China's Eco-city Construction*, Research Series
on the Chinese Dream and China's Development Path,
DOI 10.1007/978-3-662-48153-0_4

must be geared toward maximizing efficiency or effectiveness, which requires careful utilization of the limited resources.

4.1.2 Definition and Qualities of a Resource-Conserving City

Definition Current global social development indicates that resources on Earth are increasingly unable to support further development of the consumption-oriented industrial society and that resource scarcity has reached the level of global crisis. Human beings must find new economic growth and production patterns and lifestyles. Resource conservation in production, distribution, consumption, and other fields is key to future transformation because most people live in cities.

Resource-Conserving Cities: The Foundation Humanity's Survival and Development Modern Times The number of people moving to cities will continue growing in the twenty-first century. Urban population is expected to represent more than 80 % of the world's total population in 2050, and 75 % of Chinese people will live in cities. The majority of human beings will live, work, and amuse themselves in cities. The production patterns and lifestyles in cities will largely decide how human society exists and evolves because cities house most human being and their material and spiritual wealth. The traditional and inefficient urban economic development and consumption pattern have been exacerbating scarcity of energy sources, freshwater, land, minerals, and other resources in China. Resource scarcity will increasingly impede urban development and deteriorate working and living environments if no change is made to the traditional production pattern that suffers from high input and consumption and low efficiency and if no efforts are made to raise awareness of conservation and promote resource-conserving lifestyles across society. This will severely affect the survival and development of humans.

Resource-Conserving Cities Play a Leading Role in a Resource-Conserving Society A resource-conserving society advocates recycling resources, human-nature symbiosis, and eliminating non-sustainable production patterns and lifestyles. Consumers follow resource conservation and sustainable socioeconomic development practices in this society. A resource-conserving society seeks to maximize resource conservation, improve resource utilization, and minimize environmental damage as long as standard of living continues improving. Urbanization is the only path to modernization. China is gradually eliminating policy and institutional barriers to urbanization after 20 years of reform and opening up and has made great achievements in economic and urban development. Economic development and urbanization indicators show that urbanization in China has been progressing rapidly. Cities are large consumers of water, energy, land, minerals, and other resources, leaving them the most important actors in creating a resource-conserving society. A resource-conserving society must first construct resource-conserving cities.[2]

[2] Chen [2].

Resource-Conserving Cities Are the Material Basis for a Sustainable Urban Form Urban form is a complex economic and cultural phenomenon as well as a social process. It mixes urban society, economy, and culture. Urban form includes the formation of urban spaces, the spatial arrangements of human activities, the spatial arrangements of land use, the descriptions of urban landscapes, etc. Urban form reflects the past and present status of urban culture, technology, and social behavior as well as future urban development trends. Kevin Lynch said that a city can be seen as a story and an illustration that reflects the relationships in a crowd. A resource-conserving city provides for a city's existence, evolution, and development even accounting for the varying processes of urban development and urban form in different cities. The resulting conclusion is that "resource-conserving cities will make life better."[3]

Resource-Conserving Cities Are Necessary for a Harmonious Society Resource-conserving cities are necessary for sustainable development and a harmonious society. A resource-conserving city relies on two basic urban components – individuals and organizations. Every individual in the city must become aware of conservation, conserve resources in his or her daily life, and effectively improve resource utilization based on his or her experience and knowledge. A resource-conserving city is possible only if resource conservation values are instilled throughout the city. Individuals tend to work in businesses, governments, or other organizations and are generally affected by their organizations' resource conservation efforts. Individuals and organizations influence each other in resource-conserving city construction.

Resource-Conserving Cities: A Necessary Outcome of Good Governance A resource-conserving city covers all aspects of production and of people's lives in the city and will raise the municipal government's management requirements because it organizes and implements resource conservation programs. The municipal government should organize and coordinate development, create an interdepartmental coordination mechanism, clarify departmental responsibilities, and strengthen guidance, coordination, supervision, and inspection. The municipal government should also lead necessary actions. Municipal government authorities in China perennially suffer from understaffing, small budgets, and overspending. Misguided decisions and policies made by government officials that have wasted financial resources have become a major obstacle to resource-conserving city development. A resource-conserving city requires improved management of the city and the municipal government itself to create a limited, efficient, provident, and service-oriented government and to achieve good governance. A resource-conserving city will necessarily result from good governance.[4] A resource-saving city needs resource conservation prioritized over resource exploitation and economic growth pattern transformation. Governments should improve resource utilization and develop a circular economy along with measures to accelerate economic restructuring, technological progress, and improved laws and policies.

[3] Xi and Xu [3].

[4] Xi and Xu [3].

Governments should raise awareness of conservation, immediately establish and improve institutions and mechanisms conducive to building a resource-conserving city, and gradually form resource-conserving economic growth and consumption patterns. Urban sustainable development can further be advanced by efficiently using and recycling resources.

4.1.2.1 Characteristics of Resource-Conserving Cities

Resource-conserving cities involve all aspects of urban production and life. Urban structure, components of the city, urban economy, and urban civilization provide background for the following five characteristics of a resource-conserving city[5]:

Systematic A resource-conserving city relies on systematic development and uses mathematical logic to optimize its development, operating costs, and urban functions to create harmony between urban residents and the environment, resources, and society. Study of various city subsystems, such as the social, political, cultural, economic, resource, and environmental subsystems using accepted theories and methods, is needed to build such a city. Dynamically understanding correlations between these subsystems and comprehensively coordinating the status of each subsystem to maximize the city's outputs will further aid the process.

Positive Interactions Among All Components Allow for Order A resource-conserving city has conservation-focused government authorities, organizations, businesses, households, etc. The municipal government should work to achieve orderly urban development. Businesses should use their capital in appropriate production activities. Citizens should spend money on moderate consumption. A resource-conserving city can operate in an orderly manner only if all interactions between the various components are maintained positive.

Optimal Economic Structure A resource-conserving city has an economic structure that follows the Pareto principle and the circular economy model. Its production structure uses minimum input for maximum output. Its consumption structure must become green as accepted by local citizens, including widespread efforts to save energy and resources to contribute to the city's sustainable development.

Enhanced Civilization A resource-conserving city enhances urban civilization. It embraces sustainable development and promotes harmony between humanity and nature. Simplicity and frugality are admired. Love of technology is fashionable. New ideas, knowledge, and concepts continuously stream out.

Highly Developed Science and Technology The local government and relevant organizations must build public service systems, urban infrastructure, factories, and residential buildings that embrace sustainable development and a circular economy to develop toward a resource-conserving city. This leads to heavy reliance on mod-

[5] Zhao et al. [4].

ern science and technology, especially advanced technology that enables resource conservation. Resource-conserving cities advance dynamically as science and technology develops. The building and construction industry serves as a pillar for the national economy but consumes great amount of energy. Developing and introducing energy-efficient technology in the building and construction industry is important for resource-conserving city development. Information technology has become increasingly important in modern urban management and public services as urbanization and information and network systems progress, leading to digital communities and cities. Information technology is largely responsible for urban modernization and will undoubtedly be a key technological element of a resource-conserving city.[6]

4.2 Assessment of Resource-Conserving Cities

4.2.1 Developing Assessment Metrics for Resource-Conserving Cities

Objectives Resource-conserving city creation requires a comprehensive and accurate assessment of resource conservation across the city. Assessment metrics for resource-conserving cities are a prerequisite for conducting assessment work. Such metrics simplify complex relationships between the natural environment and social and economic activities and provide quantitative indicators that explain progress in resource utilization, conservation, and sustainability to the public, governments, and enterprises during urban development. These metrics reflect trends and changes in all aspects of the city, including the natural environment, society, and the economy, to measure the level of resource assurance and the potential of sustainable development in a particular region, and answering questions of general concern such as: what is the current level and future trend of sustainable development in this region? How does this level compare with other regions? What approaches can enhance the potential of conservation development? Policies and measures for improvement should be implemented to achieve coordinated development that considers economical resource use.[7]

Design Principles Assessment metrics for resource-conserving cities form a large system, as they provide the foundation for government decisions at all levels and will eventually be used in national accounts. Certain principles are necessary to develop these metrics. A region's economic and social development measurement should be scientific, practical, simple, and stable. A universally accepted set of design principles is difficult to put together because of the diversity of disciplines. Metrics that are able to objectively, comprehensively, and scientifically measure and

---

[6] Li [5].

[7] Liu and Ma [6].

evaluate the status and potential of resource conservation in a city should follow the following seven principles:

(1) Scientific. Individual resource-conserving city's assessment metrics' indicators must be conceptually clear, scientifically meaningful, and accurately reflect the objectives of building a resource-conserving city.
(2) Comprehensive. Urban development involves land, water, minerals, and other resources connected to the economy, society, ecology, and the environment. No urban resource conservation indicator set can completely cover all aspects. An indicator set should nevertheless attempt comprehensive coverage and reflect the major aspects of resource conservation and urban development.
(3) Comparable. Each individual indicator should reflect the level and potential of resource conservation in a city but should also consider universality to allow comparisons internationally, nationally, and between cities. Various indicator levels are needed to reflect the construction and management of infrastructure and public facilities in resource-conserving cities of varying sizes and functions. Indicators should comprehensively and accurately reflect the degree and potential of resource conservation and should allow for comparisons between different cities and time periods.
(4) Independence. Indicators should be independent of each other. Independence avoids repeated evaluation and calculation using the same resource utilization and conservation data and helps organize and structure the indicator system.[8]
(5) Operability. Data availability and realistic expectation should be considered when developing indicators. Indicators difficult to quantify or qualify should be as few as possible. Necessary qualitative indicators should be quantifiable when possible, and quantitative indicators should be calculable, directly or indirectly, using data released by the national statistics authority.[9]
(6) Usability. Assessment metrics for resource-conserving cities are designed to guide urban development toward a resource-conserving city, which requires that all individual indicators are easy to understand and accept for the majority of people and that they help governments, businesses, and the public learn about resource utilization problems and progress to better further improvements.
(7) Stability. Metrics will necessarily change along with socioeconomic development. Resource-conserving city's indicators should remain relatively stable within a certain period of time to facilitate comparisons and analysis on resource conservation status and trends during urban development.

Framework and Content Assessing resource conservation in a particular city requires consideration of future trends and of resource conservation potential in the city in addition to measuring the current effects of its efforts. The four layers in the indicator set for a resource-conserving city follow the resource-conserving city objec-

[8] Liu and Li et al. [7].

[9] Ye et al. [8].

tive and design principles. The resource-efficient composite index (RECI) consists of two criteria layers – the conservation status and potential indices – each containing a corresponding indicator and sub-indicator layer. The conservation status index is used to determine whether the current status of resource consumption, economic development, and industry mix in a city is in line with conservation goals. The conservation potential index is used to examine whether resource conservation trends in the city favor resource conservation development. The conservation potential index includes technical, social, and institutional support indicator layers (see Table 4.1).

Table 4.1 Assessment metrics for a resource-conserving city

Objective layer	Criteria layer	Indicator layer	Sub-indicator layer
RECI	Conservation status index	Resource consumption indicator	Economies of agglomeration index
			Population concentration index
			Percentage of unused land
			Water consumption per 1,000 yuan of GDP
			Domestic water consumption per capita
			Percentage of sewage treated and reused
			Power consumption per 1,000 yuan of GDP
			Domestic power consumption per capita
			Primary energy consumption
			Percentage of solid waste reused
		Economic development indicator	Percentage of services in GDP
			Percentage of clean energy sources used
			Percentage of culture industry
			Percentage of circular economy
			Percentage of green buildings
			Percentage of green vehicles in use
		Conservation across society indicator	Population growth
			Percentage of green government procurement
			Percentage of public transportation
			Penetration of water conservation equipment
			Percentage of residents using gas
			Penetration of centralized heating

(continued)

Table 4.1 (continued)

Objective layer	Criteria layer	Indicator layer	Sub-indicator layer
	Conservation potential index	Technical support indicator	Percentage of R&D expenses in GDP
			High-tech industry growth
			Percentage of research results yielding products
			Contribution of technological advances
		Social support indicator	Green consumption by citizens
			Citizens' values
			Corporate social responsibility (CSR)
			Average number of years in school per capita
			Number of technical professionals per 1,000 persons
		Institutional support indicator	Percentage of education expenses in GDP
			Government policy and law enforcement
			Technology administration and performance
			Technology promotion system and performance

4.2.2 *Methods for Assessing Resource-Conserving Cities*

Reasonable assessment methods are needed to calculate the status and potential of resource conservation in a city in addition to a comprehensive set of indicators. There are a variety of assessment methods, such as the Delphi method, gray correlation evaluation method, fuzzy comprehensive evaluation method, entropy method, and hierarchical analysis method. Based on research results from Zuo Qiting et al.,[10] each individual indicator for a resource-conserving city was evaluated using resource-efficient degree (RED); the status and potential of resource conservation in the city were evaluated using the conservation status and potential indices; and the overall status of resource conservation in the city was evaluated using the RECI.

[10] Zuo and Li [9].

4.3 Building Resource-Conserving Cities

4.3.1 International Comparison of Resource Consumption and Conservation

Data show vastly different energy consumption levels per unit of GDP in different countries. With the exception of Russia that has a strong stake in heavy industry, China had the highest energy consumption per unit of GDP in 2001, 4.5 greater than Germany's, 1.2 times greater than India's, 6.6 times greater than Japan's, 3.7 times greater than the United States', 2.6 times greater than Australia's, and 2.4 times greater than Brazil's. The amount of carbon dioxide produced per unit of GDP is also much higher in China than in other countries, as much as 12 times higher than in the level in Japan which enjoys the lowest emissions. Table 4.2 demonstrates that the average level of energy consumption for industrial products in China anywhere from two to four times the level in developed countries.[11]

Table 4.2 International comparison of energy consumption by industrial product

Product	Item	Energy consumption in China	Energy consumption outside China	China/ non-China
Steel	Comparable energy consumption per ton of steel (kgce/t)	727 (Baosteel was the most energy-efficient steelmaker in China in 2000)	646 (Japan in 2000)	1.13
Synthetic ammonia	Large-scale projects, with natural gas introduced (kgce/t)	1,200	970 (the United States; large-scale natural gas projects in 1997)	1.24
Caustic soda	Membrane process at major chlor-alkali makers (kgce/t)	1,600	1,250	1.28
Ethylene	Comprehensive energy consumption (kgce/t)	1,210 (1997)	870 (1997)	1.39
Cement	Comprehensive energy consumption (kgce/t)	194 (Medium and large enterprises in 1997)	123 (Japan in 2000)	1.52
Ceramics	Building (kgce/m^2)	6.7	1.5 (Italy)	4.47
	Sanitary (kgce/piece)	15.0	3.5 (Germany)	4.29
Bricks and tiles	Hoffmann kiln (kgce/1,000 pcs)	110	78 (Germany)	1.41
	Tunnel kiln (kgce/1,000 pcs)	135	45 (Germany)	3.00
Lime	(kgce/t)	160	130 (Japan)	1.23

Source: Bai Quan and Tong Qing. It is imperative to make energy-intensive industrial products more competitive by increasing energy efficiency. *Economic Survey*, March 2004

[11] Liu, Li et al. [7].

Table 4.3 International comparison of water resource utilization (2004)

Country	Water consumption per 1,000 yuan of GDP (m^3)	Effective utilization of water for agricultural irrigation	Water consumption per 1,000 yuan of industrial added value (m^3)
China	40.6	0.45	21.8
Developed countries	5.0	0.7–0.8	5.0

Source: China Statistical Yearbook 2003 and International Statistical Yearbook 2003

Table 4.4 International comparison of land use (2002)

Country	Land area (10^6 ha)	Farmland area (10^6 ha)	Percentage of farmland area out of land area (%)	GDP per unit of land area (1,000 USD/ha)
World	13,066.88	1,404.13	10.8	241
Developed countries	5,462.78	611.42	11.2	456
Developing countries	7,604.10	792.70	10.4	86
USA	915.90	176.02	19.2	1,074
Germany	34.90	11.79	33.8	5,368
UK	24.09	5.75	23.9	5,872
France	55.01	18.45	33.5	2,353
China	932.74	130.04	13.9	116
Russia	1,688.50	123.46	7.3	14.9
Indonesia	181.16	20.50	11.3	84.6
Japan	36.45	4.42	12.1	13,283
South Korea	9.87	1.68	17.1	4,631

Source: Liu X, Li X, et al [7]

Water consumption per 1,000 yuan of GDP in China is eight times the average level in developed countries. Water consumption per 1,000 yuan of industrial added value is four times the level in developed countries. Only about 50 % of water used for agricultural irrigation is effectively used (Table 4.3). The international comparison of land resource use is shown in Table 4.4.

China significantly lags behind developed countries in renewable resource recycling and comprehensive utilization of resources. Retreads in 2003 serve as an example (see Table 4.5). Retreads accounted for only 4 % of new tire production in China, compared with 10 % in developed countries and 18.8 % in the EU. An increasing number of automobiles in China will result in a growing number of used tires. These tires will waste resources and pollute the environmental if they are not recycled. China lags far behind developed countries in recycling and comprehensive utilization of mineral resources, wood, and industrial solid waste (see Table 4.6).

Table 4.5 International comparison of renewable resource recycling (2003)

Country	Percentage of reused steel in crude steel production (%)	Percentage of reused copper in copper production (%)	Percentage of reused aluminum in aluminum production (%)	Percentage of retreads in new tire production (%)
China	26	22	21	4
Developed countries	43	37	40	10

Sources: Transformation of China's economic growth pattern: achievements and problems. China Development Research Foundation (CDRF)

Table 4.6 International comparison of comprehensive resource utilization (2004)

Country	Percentage of mineral resources recovered (%)	Comprehensive wood utilization (%)	Comprehensive utilization of industrial solid waste (%)
China	30	60	55.4
Developed countries	50	80	100

Source: Liu, Li et al [7]

4.3.2 Measures for Building Resource-Conserving Cities in China

Global best practices and experience indicate that resource-conserving city construction should focus on the following issues:

Improving Legislation Experiences in other countries, especially in Japan and Germany, indicate that resource-conserving city development requires a reliable legal system. Legal improvements may include promoting resource recycling, green energy development, and other measures for resource conservation; regulating the production and consumption patterns; adjusting and optimizing industry taxes; optimizing the industry mix and guiding the economic growth pattern transformation; clarifying legal resource ownership and increasing resource utilization costs to reduce resource waste; monitoring government departments' use of public funds and management of public affairs; and improving the quality and effectiveness of urban infrastructure to avoid waste caused by damage or demolition that precedes expected dates. Japan's recycling-centered society serves as an example. Japan focuses on promoting recycling and comprehensive utilization of waste and has gradually formed a set of relevant laws along with corresponding regulations and policies drafted by government departments. They played a key role in advancing Japan to become among the best resource recycling and environment-protecting countries. Germany promulgated a law on circular economy and waste management in 1996 that centered the German legal system on a circular economy. The United

States promotes green energy development and utilization through legal means. Under the Energy Policy Act, $3.4 billion USD was allocated to local governments every year during 2004–2006 for recycling used home appliances and encouraging the purchase of new energy-efficient products.

Governments Play a Leading Role Governments should create outlines and guiding opinions for resource-conserving city construction, clarifying objectives, basic principles, and responsibilities of all members of society. These outlines and opinions are critical in providing clear orientation and code of conduct for governments, organization, and individuals in the resource-conserving city development process. Economic leverage, incentives, and restraint mechanisms are all crucial for resource-conserving city progress. Governments should promote the conservation and efficient use of resources by drafting and improving tax, pricing, financial, and other policies and measures. High taxes should be levied on energy-intensive and heavy-polluting industries and products, taxes should be lowered or eliminated, and subsidies should be offered to energy-efficient and light-polluting industries and products to economically stimulate people to save energy and reduce pollution. Obsolete production processes and equipment must be forcibly eliminated, and reasonable water, fuel, electricity, and gas pricing, repricing, and accountability mechanisms linked with the market must be established. Governments should launch resource allocation and conservation programs to reduce emissions and pollution. Resource utilization should be maximized by employing comprehensive utilization and recycling of resources in extraction, production, consumption, and waste treatment. Organizations should actively develop new energy sources and renewable resources and optimize the energy sources and resources mix. Governments at all levels should reform and improve the performance monitoring and evaluation system to match the Scientific Outlook on Development. Blind investment and redundant projects should be curbed, and resource waste and environmental damage caused by pursuing GDP growth should be avoided. Overall plans for creating resource-conserving cities and specific plans for garbage treatment, renewable resource use, and other issues are essential for progress. The urban system and spatial arrangements should be designed to avoid energy, water, land, and other resource waste that can occur when spatial arrangements are unreasonably managed.[12]

Enhancing Technology Innovation and Developing a Circular Economy Former Chinese Premier Wen Jiabao stated that creating a resource-conserving society should rely on scientific and technological progress and innovation to provide a technical support system for resource conservation. Research and development for technology critical for resource conservation and recycling should be encouraged. New technologies, processes, equipment, and materials that help resource conservation efforts and provide support for resource conservation projects and circular economy development should be promoted to achieve sustained and rapid growth in the high-tech industry and improve traditional industries and circular economy

[12] Liu, Li et al. [7].

development. The Chinese Ministry of Science and Technology (MOST) ran energy-conserving building demonstration using a variety of energy-saving technology, green technology, and comprehensive integration of smart technology to arrive at low cost and energy efficient and succeeded in securing annual savings of more than 900,000 kWh, more than 70 % reduction, creating a scientific and technological resource conservation model. An energy-efficient building demonstration was run as a cooperation between the MOST and the US Department of Energy. This inexpensive green building integrates energy-conserving, green, and intelligent technology to set a standard of technological resource conservation, featuring annual power savings of over 900,000 kWh, over 70 % energy savings per year.

Enhancing Public Awareness and Resource Conservation Habits The public's critical participation in building resource-conserving cities makes public awareness of resource conservation and skills to enact conservation and integral piece of effective resource-conserving city construction. Public participation in building resource-conserving cities must be encouraged along with a conservation atmosphere, healthy lifestyles, and sustainable consumption patterns. Relevant organizations should spread resource conservation-related knowledge and provide the public with opportunities to learn skills and techniques to conserve resources in their daily lives. TV, radio, and the Internet should be engaged to further popularize resource conservation knowledge and raise public awareness of resource crises. Resource-efficient school and community initiatives and resource conservation and waste recycling guidelines available to all people will further the cause. The goal is gradual progress toward resource-conserving and environmentally friendly lifestyles and consumption patterns.

4.4 Beijing: A Resource-Conserving City

Beijing is a large city facing severe resource shortages and high consumption. Resource scarcity has long presented a great obstacle to Beijing's socioeconomic development. Municipal government efforts toward resource conservation in recent years have led Beijing to the foremost position in resource conservation in China.

4.4.1 Building a Resource-Conserving City: Objectives, Ideas, and Measures

The Beijing Municipal Commission of Development and Reform (BJMCDR) released the *Outline of the Program for Accelerating the Development of a Circular Economy and Building Beijing into a Resource Conserving City and the Action Plan 2005* on June 6, 2005.

General Idea and Objective The municipal government departments and organizations should further socioeconomic development and improve people's standard of living in their attempts to create a world-class livable city, making Beijing a leader in resource efficiency and environmental and ecological management in China and developing the modern international metropolis to contain an optimal industry mix, high resource efficiency, a comfortable environment, sustainable development, and social harmony.

General ideas include a focus on increased overall efficiency; relying on participation by the government, businesses, and the public; drafting supportive policies and regulations and creating an eco-friendly industry mix; promoting energy conservation and resource utilization; developing renewable energy sources and green consumption; and prioritizing planning, establishing standards, drafting policies, increasing law enforcement capacity, promoting a circular economy, enhancing resource conservation, optimizing the industry mix, and increasing publicity.

The general objective of the project is to attain desired levels in specific indicators such as energy efficiency, energy consumption per 1,000 yuan of GDP, the percentages of high-quality and renewable energy sources out of total energy consumption, water consumption per 1,000 yuan of GDP, the percentage of water reused in the city, the utilization of industrial waste, and the overall MSW utilization.

4.4.1.1 Tasks and Measures

Making Relevant Plans and Standards, Improving Relevant Policies, and Enhancing General Guidance Beijing should continue creating plans such as the *Beijing Circular Economy Development Plan* and the *Beijing Medium and Long Term Energy Conservation Plan*; developing standards such as the *Indoor Temperature Monitoring Standard for Energy-efficient Air Conditioning in Public Buildings* and the *Fan Monitoring Standard for Energy-efficient Centralized Air Conditioning Systems in Public Buildings*; and improving industry, pricing, tax, investment, incentive, and other policies.

Enhancing the Legislation and Law Enforcement Capacity and Setting Up Service Teams Beijing should engage in legal research on circular economy development and energy monitoring and management; draft and improve laws relevant to water conservation, cleaner production audits, waste treatment, and renewable energy source utilization; strengthen law enforcement capacity; and establish service teams specializing in energy conservation.

Promoting the Development of Circular Economy Beijing should encourage enterprises to adopt cleaner production; select enterprises in the petrochemical, power, pharmaceutical, electroplating, and other industries for pilot cleaner production projects; and establish cleaner production audit agencies and advisory service teams for important industries.

It should further encourage extended resource utilization between enterprises, build EIPs where resources will be recycled, and promote eco-friendly rural and agricultural development.

Further steps include providing support for resource recycling across society; promoting waste reuse specialization and waste to product conversion using market based techniques; promoting separation of waste for collection and treatment; and reducing, recycling, and converting waste into harmless substances.

Efforts in Energy and Resource Conservation Beijing should optimize the energy mix and enhance energy management, increase the percentage of high-quality energy sources out of total energy consumption, strengthen energy efficiency management, and increase energy efficiency.

It should further increase land use efficiency; reduce land demand pressure; establish a land use evaluation system; develop land use project targets for site ratio, investment strength, and output ratio; and control total land supply and the mix of land supply options.

Other useful steps include reducing water consumption and accelerating water conservation efforts; enhancing water conservation in the primary, secondary, and tertiary sectors by employing pollution prevention, managing rainwater and floods, and recycling wastewater; and adhering to water quotas and consumption targets.

Improving the Industry Mix and Optimizing Resource Allocation Beijing should "optimize the primary sector, enhance the secondary sector, and expand the tertiary sector" as per the established guideline; restrict projects that will consume large quantities of energy, water, and land and cause heavy pollution; promote a circular economy in the economic development program; and create world-class enterprises.

Enhancing Publicity to Raise Public Awareness of Environmental Protection and Resource Conservation Beijing should publicize the cause through events like the annual Energy and Water Conservation Publicity Week to raise public awareness of environmental protection and resource conservation and encourage participation in resource reuse and conservation activities.

4.4.2 Building a Resource-Conserving City: Achievements

Beijing achieved marked progress in resource-conserving city development thanks to years of efforts. In 2009, Beijing's energy consumption per 1,000 yuan of GDP fell by 5.57 % year on year, which placed it as a leader in energy intensity decrease in China and as the only Chinese municipality or province to have met energy conservation targets for 4 consecutive years. Beijing's absolute energy consumption of only 0.06074 t of standard coal per 1,000 yuan of output value was the lowest across China. That translates into the highest energy efficiency in China. In 2009, Beijing's total water consumption was 3.55 billion cubic meters, an 8.12 % drop in water consumption per 1,000 yuan of GDP from 2005 levels using adjusted prices. Reports indicate that the energy and water consumptions per 1,000 yuan of GDP in Beijing dropped by 23.34 % and 33.77 %, respectively, during the first 4 years of the 11th Five-Year Plan period. These numbers represent significant achievements in energy conservation,[13] such as:

Optimized Economic Structure Beijing's economic structure is gradually optimized as the service sector, high-tech industries, advanced manufacturing, and modern agriculture experience remarkable growth. The tertiary sector accounts for over 70 % of the local economy; communications equipment, computers, and electronics manufacturing have grown 53.9 % year on year; and the rapidly growing energy-efficient, high added-value industries contribute directly to a decrease in energy consumption. Beijing has gradually established industry selection and comprehensive evaluation mechanisms to phase out energy-intensive and heavy-polluting industries and enterprises. Beijing has moved more than 100 of these resource heavy enterprise fields covering papermaking, coking, textiles, chemicals, and printing and dyeing out of its urban areas in recent years. Beijing shut down 13 shaft kiln cement production lines and seven flat glass production lines in 2006 alone. These moves are expected to reduce annual energy consumption by 207,000 t of standard coal.

A Continually Improved Resource Conservation Monitoring Mechanism The *Beijing Municipal Measures for Monitoring Energy Conservation* went into effect on July 20, 2006. The Beijing municipal government established an agency specializing in monitoring energy conservation. This agency checks total energy consumption, energy consumption per product, and energy consumption of primary user equipment and has legal standing to handle discovered problems. The local government conducts energy utilization surveys and enhances categorical measurements and information monitoring, which inserts energy conservation into the socioeconomic development evaluation system. Beijing has established a city-wide energy and water consumption bulletin system to publish the rate of energy and water consumption per 1,000 yuan of GDP and other data on a quarterly basis. It has also developed mandatory and voluntary standards for energy conservation monitoring, energy consumption restrictions, general technological principles, among other projects. The local government has introduced a series of resource conservation laws and regulations such as the *Measures for Implementing the Energy Conservation Law of the People's Republic of China in Beijing City*, the *Decision on Accelerating the Development of Circular Economy in Beijing City and Building a Resource Conserving City* and the *Measures for Conserving Water in Beijing City*.

Further Developments in New Energy Sources and New Technologies Beijing has been eagerly developing renewable energy sources such as solar, geothermal,

[13] "Beijing saw the greatest decrease in energy consumption per 1,000 yuan of GDP in China in 2009 thanks to efforts in building a resource conserving city," http://www.btv.org/btvindex/xw/content/2010-05/24/content_960702.htm, May 24, 2010.

wind, and biomass to further develop the circular economy. The Olympic Village features a 6,000 m^2 solar thermal system that provided up to 16,000 athletes and officials hot water for bathing. Just this system conserves about 2,000 t of coal per year. New and renewable energy sources equivalent to 2.6 million tons of standard coal were used in Beijing in 2010, accounting for 4 % of total annual energy consumption. Many local businesses, office buildings, and shopping malls are also actively adopting efficient wall insulation technologies, advanced heating and air conditioning systems, energy-efficient lighting control systems, and other energy-saving technologies.

Forming a Resource Conservation Atmosphere and Conservation Habits Beijing municipal government agencies and departments were the first of their kind to implement resource conservation projects. The municipal government launched the Energy-Efficient Government initiative as a top priority in 2005 and took three measures including conducting an energy consumption survey of 54 government agencies and departments; signing the Agreement on Responsibility for Energy Conservation at Government Agencies, which set an 8 % energy conservation target; and identifying ten government agencies and departments for pilot energy conservation projects. A year-end survey showed that these 54 government agencies and departments reduce energy consumption by 11 % year on year, which is equal to 2,600 t of standard coal. The energy conservation initiative covered government agencies and departments in 18 districts across Beijing by the end of 2006. Public awareness of resource conservation was enhanced through these governmental efforts. Beijing exhibits on energy conservation and environmental protection attract many attentive visitors that go on to ask questions and share experiences with energy-efficient products and technologies. Penetration of home water conservation equipment in 2006 reached 79.3 % in Beijing's urban areas and 72.5 % in its suburban districts (Table 4.7).[14]

Table 4.7 Top 50 Chinese cities by RECI, 2008–2011

2008		2009		2010		2011	
City	Ranking	City	Ranking	City	Ranking	City	Ranking
Shenzhen	1	Guangzhou	1	Guangzhou	1	Shenzhen	1
Nanjing	2	Shenzhen	2	Shenzhen	2	Guangzhou	2
Guangzhou	3	Nanjing	3	Nanjing	3	Beijing	3
Shanghai	4	Beijing	4	Beijing	4	Shanghai	4
Beijing	5	Shanghai	5	Shanghai	5	Nanjing	5
Dongguan	6	Zhuhai	6	Zhuhai	6	Hangzhou	6
Zhuhai	7	Dongguan	7	Hangzhou	7	Zhuhai	7
Xiamen	8	Xiamen	8	Xiamen	8	Xiamen	8

(continued)

[14]The MOST's official website, "Beijing steps up building a resource conserving city for higher quality of life," *Science and Technology Daily*, July 19, 2006.

Table 4.7 (continued)

2008		2009		2010		2011	
City	Ranking	City	Ranking	City	Ranking	City	Ranking
Hangzhou	9	Hangzhou	9	Xi'an	9	Xi'an	9
Xi'an	10	Hohhot	10	Yulin	10	Hulunbuir	10
Jinan	11	Baotou	11	Wuhan	11	Dongguan	11
Dalian	12	Jinan	12	Dalian	12	Sanya	12
Wuxi	13	Dalian	13	Jinan	13	Dalian	13
Wuhan	14	Langfang	14	Sanya	14	Tianjin	14
Ningbo	15	Tianjin	15	Nanning	15	Haikou	15
Haikou	16	Wuhan	16	Langfang	16	Jinan	16
Nanning	17	Changsha	17	Ganzhou	17	Shenyang	17
Tianjin	18	Nanning	18	Dongguan	18	Wuhan	18
Hohhot	19	Sanya	19	Tianjin	19	Ningbo	19
Huangshan	20	Heyuan	20	Changchun	20	Wuxi	20
Chengdu	21	Wuxi	21	Heyuan	21	Yulin	21
Shenyang	22	Nanchang	22	Changsha	22	Yinchuan	22
Zhengzhou	23	Haikou	23	Haikou	23	Changsha	23
Nanchang	24	Yulin	24	Shenyang	24	Changchun	24
Suzhou	25	Suzhou	25	Hohhot	25	Yantai	25
Changchun	26	Ganzhou	26	Nanchang	26	Chengdu	26
Qingdao	27	Xi'an	27	Huangshan	27	Hohhot	27
Changsha	28	Qingdao	28	Hefei	28	Qingdao	28
Langfang	29	Ningbo	29	Wuxi	29	Huangshan	29
Hefei	30	Changchun	30	Qingdao	30	Nanchang	30
Dongying	31	Chengdu	31	Yinchuan	31	Hefei	31
Sanya	32	Fuzhou	32	Ningbo	32	Suzhou	32
Lijiang	33	Taiyuan	33	Chengdu	33	Jiaxing	33
Fuzhou	34	Huangshan	34	Weihai	34	Zhoushan	34
Yulin	35	Jiaxing	35	Fuzhou	35	Zhenjiang	35
Yinchuan	36	Hefei	36	Suzhou	36	Weihai	36
Yantai	37	Dongying	37	Zhenjiang	37	Fuzhou	37
Taiyuan	38	Weihai	38	Zhangjiajie	38	Lijiang	38
Hanzhong	39	Zhongshan	39	Yantai	39	Baotou	39
Zhongshan	40	Yantai	40	Lijiang	40	Ganzhou	40
Baotou	41	Yinchuan	41	Hulunbuir	41	Harbin	41
Weihai	42	Zhenjiang	42	Dongying	42	Langfang	42
Zhenjiang	43	Lijiang	43	Taiyuan	43	Zhengzhou	43
Jiaxing	44	Zhoushan	44	Harbin	44	Nanning	44
Urumqi	45	Tai'an	45	Zhongshan	45	Tai'an	45
Chongqing	46	Yichun	46	Lianyungang	46	Shaoxing	46
Zhoushan	47	Baoding	47	Baotou	47	Yichun	47
Ganzhou	48	Zhengzhou	48	Jiaxing	48	Taiyuan	48
Harbin	49	Lianyungang	49	Tai'an	49	Dongying	49
Tai'an	50	Hanzhong	50	Zhengzhou	50	Zhangjiajie	50

References

1. Huang K (2009) Analysis of reasons for choosing a certain approach to building a resource-conserving or environmentally friendly city. Rule Law Econ (6):126–127
2. Chen D (2005) Preliminary research on the basic meaning of a resource conserving society. China Popul Resour Environ (2):5–9
3. Xi J, Xu B (2005) Resource-efficient city. China Modern Economic Publishing House, Beijing
4. Zhao J, Guo Q, Lv W (2008) Research report on building a resource conserving society in China: theories and practices. China Modern Economic Publishing House, Beijing
5. Li N (2006) Resource conserving cities: the key to building a resource conserving society. Theory Front (3)
6. Liu R, Ma Y (2007).On integrated assessment metrics for building a resource conserving society. J Shandong Univ Finance (4):54–57
7. Liu X, Li X et al (2010) On resource conserving agriculture and cities. Beijing Normal University Publishing Group, Beijing
8. Ye W, Yu Z, Tang J (2004) Preliminary study on how to develop metrics for a resource conserving society. Coal Econ Res (11):8–9
9. Zuo Q, Wang L (2008) Resource-saving society: evaluation methods and applications. Res Sci (3):409–414

Chapter 5
Green Industrial Cities

Yongli Zhang and Jiaxing Lu

5.1 Green Industry City: Concept and Definition

5.1.1 Definition and Characteristics of Green Industry

Definition The term green industry derives from the Green Plan introduced in 1989 by the then Canadian Minister of the Environment. The plan intended coordination between economic development and environmental protection and integration of eco-friendly values into socioeconomic development by injecting the green idea the macroscopic socioeconomic development plan. The Green Plan was adopted by 12 industrialized countries in the early 1990s and developing green industries became an important strategy in countries that wanted to achieve socioeconomic sustainability because of their important in implementing the Green Plan. A series of sustainability oriented concepts emerged later on, including green industry, environmental industry, environmental protection industry, cleaner production, green product, and green marketing. The definition of green industry has continued evolving throughout this process. The breadth, stratification, and changing nature of the green industry concept has prevented a unified definition, leaving it ambiguous in both intension and extension. It is further ambiguous because it overlaps with many similar concepts and ideas.

Y. Zhang (✉)
Business College, Northwest Normal University, Lanzhou, Gansu, China
e-mail: 894816361@qq.coml

J. Lu
Lanzhou City University, Lanzhou, Gansu, China
e-mail: 1291106856@qq.com

J. Li, T. Yang (eds.), *China's Eco-city Construction*, Research Series on the Chinese Dream and China's Development Path,
DOI 10.1007/978-3-662-48153-0_5

Different countries and international organizations and bodies understand the concept differently based on their individual needs and characteristics.

(1) The International Green Industry Union (IGIU) holds that if an industry attempts to conserve resources and reduce pollution out of environmental consideration using technology and green production mechanisms in the production process, it can be called a green industry.
(2) The UNDP defines green industry as products, equipment, services, and technologies that prevent and reduce pollution, such as solar, geothermal, and wind energy, public transportation, as well as other products, equipment, services, and technologies that save energy and resource inputs and increase efficiency.
(3) Green industry is defined differently due to the diversity of green industries and of each country's purposes. Green industry is called environmental industry in the United States and eco-friendly industry or eco-friendly business in Japan. Terminology aside, green industry serves as a generic term for technology and product development, distribution, resource utilization, information services, project contracting, and other activities economic activities meant to prevent and control environmental pollution, improve the environment, and protect natural resources.

Characteristics Green industry is globally eco-friendly, implements recycling, spreads through the industrial system, use interdisciplinary angles, and works for the common good. Despite continuous changes and development around the world, green industry has demonstrated some basic common characteristics, including:

Global Acceptance and Unified Standards Unified product standards and standardized production processes are necessary for green industry development. The International Organization for Standardization (ISO) has released the ISO 14001 Environmental Management Systems as a follow up to the ISO 9000 quality management series, which covers environmental labeling as an important component. The ISO is working on globally unified environmental product labeling. The IGIU is a UN-led international NGO and one of the world's most authoritative and influential green industry organizations. It launched IGIU Green Product as a professional certification program, so that an "IGIU" label marks green products certified by this program.

Systematic Production Developing green industry requires not systematic ideas and concepts and implementation of these ideas and concepts throughout the production process. Environmental protection must be emphasized in all steps ranging from product design through product consumption and waste treatment, and must be incorporated into corresponding environmental management practices. Green concepts must be integrated with technologies. Green industry development extends beyond any single industry, sector, or organization to a whole multitude of industries and sectors. The relationships between involved industries and sectors must be well coordinated to allow a clear division of labor, active cooperation, equality, and mutual benefits. Green industry arrangements and plans must consider regional conditions in order to achieve positive environmental and socioeconomic benefits.

Sustainable Resource Utilization Enterprises in green industry engage in cleaner production by adopting production processes that minimize waste and fully utilize resources to protect the environment. Green industry players endeavor for simultaneous socioeconomic and environmental benefits, and coordinate environmental and economic development while keeping their focus on sustainable development. Sustainable resource development and utilization are at the core of green industry.

High-Tech Production Processes Green industry philosophical developments indicate that advanced technologies and methods must be adopted in all production processes to ensure processes and products are truly green and to provide healthy and safe consumer goods. Prioritizing the developing of advanced technologies is pivotal to green industry development. This includes improving traditional processes or technologies and organically combing traditional processes with advanced technologies.

5.1.2 Definition and Characteristics of a Green Industry City

"Better city, better life" was the theme of Expo 2010 Shanghai China and has become every city's development criterion. Low carbon development, energy conservation, convenience, environmental friendliness, and harmony among human beings and between human beings and nature have long been advocated in modern cities and are now guiding urban development across China. As these green concepts become increasingly popular, they drastically shift the urban development pattern to incorporate green industry cities.

Definition A green industry city organically combines urban and rural areas, promotes green concepts, and use advanced technologies to establish a green industrial system to achieve sustainable socioeconomic and environmental development, ecological balance, and harmony between humanity and nature. A green industry city has many green functions, a high percentage of green industry in the local economy, a green and harmonious environment, advanced public services, and unified and harmonious management of socioeconomic, ecological, and infrastructure systems. Green industry urban development adheres to scientific and humanitarian principles. The following perspectives clarify the green industry city concept. (1) A green industry city focuses on humans' need for a green environment and overall human development, emphasizing people throughout development and human collaboration toward prosperity and happiness. Green industry cities will form the primary human habitat in the future harmonious society. (2) Green industry sectors represent a growing share of production in a green industry city and play a dominant role in the city's economic structure. Green tourism, technology, culture, products, production, and environment influence and determine the city's functional position and the direction of urban development. (3) A green industry city should be livable and harmonious, attempt to achieve symbiosis between humans, nature, and society, progress urban culture, further environmentally friendly actions, and achieve an eco-civilization.

5.1.2.1 Characteristics of a Green City Include

Clean and Efficient Production A green city ensures sustained economic growth that is healthy. A green industry city requires rational and efficient industry, energy, technology and product mixes, as well as cleaner production methods achieved using technologies that save energy, recycle resources, and prevent pollution. A green industry city should develop high-tech industries, the culture industry, and modern services to optimize the industry mix. It should adjust production, distribution, and consumption to effectively use resources and energy and to coordinate urban economic systems and ecosystems development in order to form a virtuous cycle.

Green and Healthy Consumption A green industry city advocates healthy and moderated consumption and promotes ethical consumerism. It transforms the production pattern by changing consumption patterns to optimize the industry mix and the economic structure and to promote green industry growth. A green industry city encourages consumers to choose green products that are uncontaminated and support public health. It asks consumers to dispose of waste properly to prevent environmental pollution. It also guides consumers to respect nature and heed health, environmental protection, and resource conservation.

Ecological Awareness and Environmentally Friendly Values Going green is a new development concept that goes far beyond nature to cover the economy, society, culture, and life. A green industry city demands that residents develop ecological awareness and environmentally friendly values as well as a sense of social responsibility and health awareness. A green industry city incorporates quality of life and population quality and health commensurate with the level of socioeconomic development. A green industry city must create a comfortable living environment, stable social order, open and democratic politics, a reliable social security system, comprehensive cultural development, green community, and eco-friendly urban spaces.

Life Harmonious with Nature Harmonious urban life goes beyond the relationship between humans and nature to include the relationships among humans. People must return to nature, respect and restore it. Love, mutual assistance, and harmony should thrive in human relationships. A good cultural atmosphere must be created through a healthy and energetic urban culture as well as the city's cultural functions, character, and charm. A green industry city should establish an advanced culture to lead industrial development and clean, efficient, healthy, equal, and harmonious urban social development.

A Garden Environment Going green and creating an eco-friendly city where humans are in harmony with nature has become increasingly popular in urban development as globalization expands. A green city controls all pollution, efficiently uses resources, and nurtures harmony between humans and nature. A green industry city is required for good environmental quality and a large carrying capacity adequate to absorb various pollutants and wastes generated by human activities.

5.1.2.2 Building a Green Industry City: Background and Reasons

With the advent of post-industrial society and the rapid development of information technologies, urban development progressed away from the industrial age to develop new trends and characteristics.

(1) An increasingly light and service-oriented industry mix has become mainstream. As urbanization approaches the intermediate stage of industrialization it exhibits a rapid development of the tertiary sector – many environmentally friendly service industries emerge along with high-tech growth. Information technology development and the IT industry rise are particularly prominent in softening the industry mix in modern cities. A technology intensive service sector is increasingly influential in urban economic development and determines a city's status.
(2) An integrated urban transportation network has become increasingly common in modern cities. Some international economic centers are currently building an efficient and low-polluting urban transportation network to maximize energy savings. Subways and light rails have played a dominant role in urban public transportation development by providing high capacity intracity passenger transportation. Railroads provides intercity passenger transportation services An urban transportation network must be safe, environmentally friendly, convenient, and efficient to promote green industry urban development.
(3) Garden cities have also become increasingly popular. A garden city aims to harmonize human's relationship with nature by employing sustainable development, ecological balance, and environmental protection to unite social, economic, and natural ecosystems and create human habitats that residents are satisfied with and that are economically efficient and ecologically balanced. Countries and governments have been taking various measures to establish garden cities. A garden city is picturesque, provides a harmonious atmosphere for human life, and also raises requirements for industrial development, promotes innovation and development of green industry while continuously highlighting and enhancing the city's character. Unique urban spaces and living environments with garden features underscore the charm typical of a given city.
(4) Urban spaces are evolving in two directions as rent increases. Commerce, services, and residential areas are gradually concentrating into central urban areas while manufacturing is gradually moving out. Development plans for central urban areas emphasize functional adjustments and the efficient utilization of urban land. The redevelopment of central urban areas in the United States made

use of seven planning and development strategies, including more pedestrian streets, renovated indoor shopping centers, preserved historic sites, developed waterfront areas, construction of office buildings and spaces for major activities, and increased transportation capacity. Redevelopment of central urban areas changes spatial structure and alters the city's industry mix.

(5) Information technology has enabled innovation in modern urban management concepts and methods. Many worldwide cities are seeking to become digital. In the United States about 50 cities are developing as digital cities. Hong Kong has started building a digital harbor. Singapore was first to develop a smart city and has continues running initiatives for this development. The increasing of information technology in cities has modernized urban management methods, lifestyles, and production methods and rendered them more efficient. The post-industrial society and the rapid development of information technology have changed urban development concepts and patterns to refocus on people and on harmony between humanity and nature. A modern city should further industrial development but also create a beautiful and comfortable living environment for residents. Urban development has modernized and now seeks to develop green cities.

5.2 Assessment of Green Industry Cities

A green industry city must have a beautiful and pleasant living environment as well as a green industry mix that supports its sustainability. The assessment metrics for green industry cities need to cover ecological, economic, social, and other factors, and must have an accurate understanding of a green city and its functions in with regard to ecosystem services and ecological and socioeconomic benefits.

5.2.1 Principles for Developing Assessment Metrics

The assessment metrics for green industry cities and the resulting model should be standard, representative, and systematic and should provide support for descriptions, explanations, assessments, early warnings, and decisions. The assessment metrics for green industry cities should follow the following principles:

Scientific Indicators should rely on scientific research and have a clear physical meaning, a standardized calculation, and statistical methods. They should objectively and truthfully reflect the primary objectives of modern urban development and the extent to which these objectives have been achieved.

Systematic The metrics must comprehensively and systematically reflect the characteristics and states of the measured elements of a green city and how these elements relate to each other, change, and develop. Specific indicators should complement each other and fully represent the green city as a harmonious, single ecosystem whose elements relate and interact with each other.

Hierarchy A green industry city is a massive complex project that requires metrics designed to comprehensively evaluate cities of such magnitude using a clear and reasonable hierarchy. Specific indicators should reflect or identify attributes of a green city on different scales or at different levels.

Comparability Indicators must be comparable along space and time. The assessment metrics should compare sustainability achievements during different time periods and demonstrate sustainability trends. They should also compare the sustainability of different regions or urban functional zones to highlight differences.

Regionality Regional socioeconomic development levels vary greatly given interregional differences in natural conditions, development, cultural background, and geographical location. Regionality should influence green city development and assessment.

5.2.2 *Metrics Makeup*

The metrics set for assessing green industry cities was developed using the principles discussed in the last section, and is shown in Table 5.1.

Table 5.1 Assessment metrics for green industry cities

	Level 1 indicator	Level 2 indicator	Level 3 indicator
Assessment metrics of green industry cities	Industrial development indicators	Economic strength	Per capita GDP
			Per capita income
			GDP growth rate
		Green industry level	The percentage of the added value of equipment manufacturing in GDP
			The percentage of the added value of culture industry in GDP
			The percentage of the added value of IT industry in GDP
			The percentage of the added value of the tertiary sector in GDP
		Green industry potential	The percentage of R&D expenses in GDP
			The size of R&D staff
			The growth rate of the environmental protection industry
			The growth rate of the tertiary sector

(continued)

Table 5.1 (continued)

	Level 1 indicator	Level 2 indicator	Level 3 indicator
	Environmental Indicators	Overall resource utilization	Energy consumption per unit of industrial added value (tons of standard coal/1,000 yuan)
			The percentage of industrial water reused (%)
			The percentage of industrial solid waste treated and reused (%)
			The overall resource utilization rate (%)
			The percentage of hazardous waste treated and disposed of (%)
		Pollution control	The amount of industry pollutants emitted and discharged per unit of industrial added value
			The amount of SO_2 emitted per unit of industrial added value
			The amount of industry pollutants emitted and discharged in a compliant manner
			The percentage of waste gathered and treated (%)
		Living environment	Per capita floor space
			Per capita area of public space
			Vegetation coverage
	Social development indicators	Green consumption and investment	The percentage of environmental investment in GDP
			The percentage of supporters of green consumption in the public
			The percentage of green government procurement
		Public behavior	Public acceptance of green products
			Public satisfaction with the urban environment
		Standard of living	Per capita disposable income
			Engel coefficient

5.2.3 *Measures for Building Green Industry Cities*

Governments Lead Green Industry Development in Cities It is necessary to clarify government responsibilities regarding green industry development. Governments must lead green industry growth by establishing corresponding organizations and bodies. The municipal government should establish an organization to lead comprehensive green industry development in order to centralize management. This organization's primary functions should include: planning and coordinating urban green industry development that is unified and market based to prevent redundant projects and resource waste; developing a reliable and comprehensive green labeling system and combining green and economic targets in criteria for assessing local government performance and the economic efficiency of local enterprises; conducting various publicity and education activities to promote green products and technologies among businesses; and advocating green consumption across society.

Governments should regulate the development of green industries. Governments must provide appropriate policy guidance along with institutional innovation and incentives to fuel green industry growth. They should establish and improve health and quality monitoring systems for green products, assist enterprises and other organizations to enhance green technology innovation, and establish an integrated mechanism for socioeconomic development and environmental protection. Governments should also reform the investment and financing systems to facilitate green industry growth, and further improve the land use and PES systems

It is necessary to develop a green market system. Distributors require help in establishing a green market system and outlets. International markets should be expanded by promoting goods and services trading as well as international economic and technological cooperation. Governments at all levels should research markets and provide information services to guide industrial and agricultural production activities with accurate market information.

It Is Necessary to Create Green Industry Development Strategies and a Green Industry System A city-specific green industry development strategy should be defined. A green industry development strategy should rely on resource utilization planning and management and avoid short-sighted development patterns that waste resources. The green industry development strategy should focus on the following points: significantly reducing environmental pollution and implementing end-of-pipe pollution control; granting green products and foods precedence over clean technology, which in turn takes precedence over end-of-pipe pollution control;

developing green products and technologies using market demand, favorable economic benefits, adequate resources, and regional advantages.

It is necessary to build a green industry system in the city. A green industry city must shift away from the traditional industry mix toward a green industry system led by comprehensive resource utilization, environmental protection, and high-tech and eco-friendly industries such as IT and other knowledge-intensive industries. Governments should develop eco-agriculture that provides green, organic food and is commensurate with local resources and the local environment's carrying capacity. Circular economy development and efficient use of natural resources are also important to push the secondary sector to go green. Green services enabled by information technology should be expanded along with general green development knowledge. Modern technology and management methods can modernize traditional services.

Green Enterprises Lead Green Industry Development Companies are leaders in green industry development and should be actively engaged in green strategy formulation and implementation. Green business development strategies and corporate green development mechanisms are needed in this process. A series of activities related to the business development strategy should focus on going green. A corporate green development mechanism using market principles is needed to promote green business development strategy implementation. Green technology innovation must be strengthened along with cleaner production implementation. Implementing cleaner production is the key to green business development. Companies should develop cleaner production programs, use green processes, organize internal material circulation, and implement measures such as process-wide control and end-of-pipe treatment to achieve green production. Green marketing and management is necessary along with bolstered green consumption. Companies should base their marketing activities on meeting consumer demand, but also on reducing resource consumption and environmental pollution. This strategy forces companies to carefully study green market demand and green consumer psychology and to conduct green product certification and pricing in order to promote green consumption.

It Is Necessary to Improve Laws and Policies to Provide Legal Support for Green Industry The law is the most important and effective external compulsory force to urge companies to advance green technology innovation, production, and marketing in a market economy. Government authorities should strengthen environmental legislation and enforcement. Government authorities should focus on prevention but enhance supervision and management in order to shift away from current Chinese ill-conceived legislation regarding green industry, unreasonably weak punishments, and lax enforcement. The green product certification system also requires improvements. China's green product certification system can be improved by aligning it with international standards such as green product standards and green product certification systems in other countries.

It Is Necessary to Enhance Green Education and a Green Culture to Raise Public Awareness of Going Green Cities need more publicity and education about the green environment. Education measures include: building a number of green schools and ecological education centers; providing education about green culture in all elementary schools, middle schools, and high schools as well as at universities; implementing educational programs for the general public that discuss protecting a green environment; providing basic, professional, and public education and job training concerning green ecology; and organically combining education of environmental ethics with going green campaigns in administration, economy, legislation, and science to establish a green culture across society and enhance people's awareness of going green.

5.3 Hangzhou: A Green Industry City

5.3.1 Background

Hangzhou is a coastal city in southeast China that sits south of the Yangtze River Delta, in the lower reaches of the Qiantang River, and west of the Hangzhou Bay. It is one of the central cities in the Yangtze River Delta region and is an important transportation hub in southeastern China. Hangzhou's exemplary natural conditions and unique environment have won it praise as a "paradise on earth" since ancient times. Hangzhou proposed an eco-city development strategy in the beginning of the twenty-first century and has since achieved impressive results that were recognized by a series of national and international awards including Garden City, Eastern Leisure Destination, Best Tourism City of China, Chinese Mainland City with the Best International Image, and one of China's Top 10 Happiest Cities. The discussions examine Hangzhou's practices, experiences, and creative ideas on green industry city development.

5.3.2 Major Steps in Green Industry City Construction

1. Practices to help optimize development deployment include: implementing an MFOZ plan based on the local environment's carrying capacity to arrange population distribution, economic deployment, land use, and the urbanization pattern; transferring industries, in particular the manufacturing industry, into the five districts (cities) outside the central urban area and accelerating the districts (cities) urbanization; promoting industry cluster development as to optimize enterprise distribution.
2. Steps to support industrial restructuring include: adopting a service first strategy whereby local government departments and organizations develop production services to encourage industry clusters and aggressively develop convenient consumer services and public services to further equality of access. Guidance and policy support for strategic emerging industries needs reinforcement. Industries in which local companies have comparative advantages, such as new-generation information technology, advanced equipment manufacturing, biotechnology, energy conservation, and environmental protection, are identified as pillar indus-

tries. Industries with great growth potential, such as new energies and materials and new energy vehicles, are identified as leading industries. High-tech industries, especially information technology, transform traditional industries. Modernizing the agricultural sector to improve yield, quality, efficiency, eco-friendliness, and safety is another necessary step along with the need to maximize the unique effects this sector has on pollution reduction and vegetation increase.

3. Other important steps include improving energy conservation and reducing emissions. Better energy conservation target accountability is necessary to achieve these goals. Local government authorities must continually improve energy conservation and emission reduction incentives, statistics collection, and supervision systems, implement energy conservation assessment and review systems for investment projects, and eliminate obsolete production technology especially in energy intensive industries. Major energy conservation initiatives are important for further progress. Local authorities and organizations must promote advanced energy conserving technology and products, cleaner production, and EPC. Expanding energy conservation programs in the industrial sector should spread to transportation, construction, and household appliances. Pollutant discharge and emission compliance of primary polluters should be increased along with tight control over the total amount of major pollutants.
4. Developing a circular economy will help circulate materials. Financial policy support and guidance can enhance resource utilization. The EPR system fuels circular economy development in production, distribution, and consumption. The industrial and agricultural sectors are the primary targets. Comprehensive mineral resource utilization is another key step. The local government encourages industrial waste recycling and promotes a renewable resource recovery system.
5. Hangzhou must develop clean energy sources. The first step to clean energy development is to control total energy consumption. Adjusting the energy consumption mix and increasing energy efficiency will help in this process. More new energy sources and clean and renewable energy sources are needed. Measures have been taken to enhance the diversification and sustainability of energy sources.
6. Progress depends on resource conservation. Resource conservation must therefore be prioritized. Comprehensive measures control total resource consumption and implement supply and demand regulations and differentiated management. Land and water resource planning, management, development, and utilization are regulated more closely.

5.3.3 Hangzhou's Green Industry City Construction Results

The results of Hangzhou's green industry city development efforts include noticeable growth of the eco-economy, environmental improvements, and an eco-friendly culture.

1. *Eco-economy*. Rapid growth of the eco-economy is noted through several indicators. The local industry mix is much more efficient. The local government promoted economic restructuring and modernized the agricultural, industrial, and service sectors as well as the culture and creative industries to modernize the overall economic system. The result is that the ratio of primary to secondary and tertiary sectors changed from 2005 where the primary sector made up 5 %, the secondary 50.8 %, and the tertiary 44.2 % to the new ration where the primary sector makes up 3.5 %, the secondary 47.8 %, and the tertiary 48.7 % in 2010. As the percentage of industry in the primary and secondary sectors declines, the tertiary sector grew significantly, further optimizing the industry mix. Technological innovation capacity has significantly advanced. Entrepreneurial activities are booming in Hangzhou, led predominantly by people with a high education level, placing the city among China's best in terms of technological innovation. Local individuals and organizations submitted 15,507 patent applications in 2009. A total of 482,215 technical professionals worked in local industries that have experienced a massive technological innovation capacity increase. The percentage of green industry has increased significantly. Local modern services industrial added value reached 289.339 billion yuan in 2010 and accounted for 48.7 % of the local GDP, compared with 44.2 % in 2005. The service sector has become the primary driving force for local economic growth and the leading employer in Hangzhou. The culture and creative industries have grown rapidly, reaching an industrial added value of 70.2 billion yuan and accounting for 11.8 % of the local GDP in 2010.
2. *Environmental improvements*. Resource consumption and environmental pollution have decreased. Combined energy consumption per 1,000 yuan of GDP dropped by more than 18.1 % during 2006–2010. Sulfur dioxide emission and industrial wastewater discharge compliance rates reached 99.5 % and 98 % respectively in 2010. Other benchmarks include the percentage of treated urban sewage, which reached 92 % and the percentage of MSW treated, which reached 100 %. Of all monitored sections in major river systems, 73.2 % tested at a water quality Level 3 or better. The percentage of good ambient air quality reached 89.6 %. An urban forest system for ecological, commercial, and landscaping purposes has greatly advanced as well as integration of the road network with the forest and river systems. Hangzhou's forest coverage was at 64.44 % in 2010, its per capita park and green space area was 15.48 square meters in urban areas, and vegetation coverage stood at 39.94 % in developed areas. Comprehensive environmental protection programs for the West Lake, Xixi Wetland Park, and the Grand Canal have been implemented. Protective and rehabilitative measures for the Qiantang, Fuchun, and Xin'an Rivers and the Qiandao Lake have been enhanced to ensure Hangzhou's distinctive river and lake landscapes. The West Lake has an average water transparency reading of 81.2 cm, the best water quality in the past 50 years. The Qiandao Lake has an overall water quality that meets Level 1 national surface water standards and its average water transparency reading is at least 7 m.

Table 5.2 Top 50 chinese green industrial cities by overall index

City	Score	Ranking	City	Score	Ranking
Beijing	65.52	1	Shenyang	38.63	26
Shanghai	63.43	2	Fuzhou	38.28	27
Shenzhen	61.79	3	Changsha	38.2	28
Guangzhou	56.91	4	Zhongshan	37.71	29
Zhuhai	50.49	5	Ningbo	37.45	30
Hangzhou	49.32	6	Weihai	36.88	31
Xiamen	48.62	7	Yantai	36.54	32
Nanjing	46.89	8	Taiyuan	36.45	33
Suzhou	46.12	9	Haikou	36.36	34
Wuxi	45.9	10	Huzhou	36.32	35
Sanya	43.21	11	Jiaxing	35.89	36
Zhenjiang	42.8	12	Zhoushan	35.59	37
Xi'an	42.53	13	Shaoxing	35.49	38
Qingdao	41.69	14	Karamay	35.25	39
Chengdu	41.56	15	Yinchuan	34.99	40
Changzhou	40.96	16	Zhengzhou	34.77	41
Dongying	40.53	17	Anshan	33.09	42
Tianjin	40.17	18	Baotou	33.08	43
Dalian	39.94	19	Tai'an	33	44
Wuhan	39.92	20	Heyuan	32.91	45
Hefei	39.74	21	Guiyang	32.35	46
Changchun	39.64	22	Lianyungang	32.26	47
Hohhot	39.38	23	Qinhuangdao	32.06	48
Jinan	39.36	24	Nanning	32.04	49
Nanchang	38.64	25	Baoding	31.92	50

3. *An eco-friendly culture*. Hangzhou has been developing into an eco-city and encouraging citizens to treat nature with gratitude, reverence, and care and to behave ecologically using both traditional Chinese cultural values that are eco-friendly and modern ecological concepts. Environmental protection, respect for nature, and green consumption are leading lifestyle choices in Hangzhou. Hangzhou has a free bike system with more than 2,000 service stations that contain over 50,000 bicycles, which are used by an average of 225,000 people per day. The project demonstrates local residents' acceptance of green travel options (Table 5.2).

Chapter 6
Circular Economy Cities

Guoquan Qian and Cuiyun Wang

6.1 Concept and Definition of a Circular Economy City

6.1.1 Concept and Characteristics of Circular Economy

Circulation has long been present in China through the ancient theory of the five elements.[1] The newer circular economy concept can be traced back to the rise of environmentalism. *The Economics of the Coming Spaceship Earth* written by American economist Kenneth Boulding in 1966 presents early circular economy theories. Boulding believed that "the earth has become a single spaceship, without unlimited reservoirs of anything, either for extraction or for pollution, and in which, therefore, man must find his place in a cyclical ecological system." The theory sees pollution as residual resources that have not been used rationally. Since resource and environmental problems are caused by development, they must be resolved by replacing the traditional economic development pattern with a circular economic pattern to protect the Earth from destruction.[2] The circular economy theory prompted earlier research on environmental resource issues, but various obstacles prevented significant progress at the time. It was not until the 1980s and 1990s that the circular economy concept finally garnered enough attention from governments and the public as global environmental problems intensified and sustainable development grew

[1] Jiang [1].

[2] Zhu [2].

G. Qian (✉)
Gansu Research Institute for City Development, Lanzhou, China
e-mail: hexiqgq@163.com

C. Wang
Lanzhou City University, Lanzhou, China
e-mail: 59174236@qq.com

J. Li, T. Yang (eds.), *China's Eco-city Construction*, Research Series on the Chinese Dream and China's Development Path,
DOI 10.1007/978-3-662-48153-0_6

in popularity.[3] British environmental economists David Pearce and Kerry Turner first formally used the term "circular economy" in the book *Economics of Natural Resources and the Environment* in 1990. Germany drafted circular economy and waste management legislation in 1996, which was the first use of the term "circular economy" in national legal texts.[4] Some developed countries have created a new, circular economy development pattern to increase environmental benefits and reduce environmental pollution, which includes planning industrial development in accordance with ecological theories. This pattern has seen considerable success in Germany, the United States, Japan, and other developed countries, where a circular economy has gradually increased in popularity in the recent years. Most developed countries have regarded a circular economy as an important approach to sustainable development. A growing number of government officials, scholars, and entrepreneurs are studying circular economy theories, offering increasingly diverse and mature options.

6.1.2 Concept and Characteristics of Circular Economy City

Concept The relationship between the circular economy and sustainable urban development can be traced back to the first mention of sustainable development. The term "sustainable development" was coined by the Brundtland Commission and was first recognized at the 1992 Earth Summit in Rio. Sustainable development is "development that meets the needs of the present without compromising the ability of future generations to meet their own needs." The World Summit on Sustainable Development (WSSD) held ten years later in Johannesburg in 2002 interpreted sustainable development in a more straightforward manner. Sustainable development is simply development supported by the three pillars of economic development, environmental protection, and social development, which are interconnected and are dependent on each other. These three pillars correspond to the economic, social, and resource environment subsystems that cover all activities and actions in the greater urban system. Coordinating these three subsystems requires coordinating between people and between humanity and nature. Sustainable urban development is a long-term and dynamic process that attempts different levels of coordination among the three pillars during different phases.

The definitions of a circular economy and an eco-city reveal that a circular economy uses the relationship between humans and nature present in eco-cities and handles conflicts between the urban economic system and ecosystems. Changes to the material flow pattern in the production and consumption processes and increased efficiency can alter a city's economic development pattern, reduce resource and

---

[3] Zhang et al. [3].

[4] Xu [4].

energy consumption, increase economic output, and reduce pollution, thereby resolving resource shortages and environmental pollution concerns. The circular economy adheres to the 3 Rs, particularly reduction, to play an important role in conserving resources and energy and in increasing efficiency of urban infrastructure, especially public resources, energy facilities, and buildings.

The circular economy is an advanced development concept and offers a new economic development pattern that values eco-economy. Circular economy applied to eco-cities provides an entirely new urban development pattern, the circular economy city. These eco-cities have reasonable spatial arrangements; reliable infrastructure; a clean and beautiful environment; high levels of safety and comfort in life; efficient material and energy use; sustainable resource use; a high degree of harmony in economic development, social development, and environmental protection; as well as symbiosis between humans and nature, rendering these cities havens for business development and living. Building circular economy eco-cities has only begun recently. Eco-city programs in China are mostly carried out in economically underdeveloped cities that suffer from long established problems and lacking infrastructure needed for future urban development. Guiyang was the first pilot circular economy city identified by the former SEPA. The advanced and new circular economy integrates clean environment, comprehensive resource utilization, and eco-design. Ecological principles guide human activities and coordinate economic growth, environmental protection, and social development. Circular economy city construction centers on three things. First, the local government should develop a moderately prosperous society. People's standard of living should improve and a beautiful environment should be available as sustained and rapid economic growth continues. Second, it is necessary to change the production and consumption patterns, gradually replacing the traditional resource consumption urban development pattern with a sustainable pattern based on a circular economy. After the economy reaches a certain size, it must reduce its reliance on resource consumption. The local government should create an environment conducive to green consumption by instituting the according policies and regulations. It should also build systems to provide environmentally friendly goods and services to shift the consumption pattern. Third, circular economy industrial systems (involving the primary, secondary, and tertiary sectors), urban infrastructure (focusing on water, energy source, and solid waste recycling systems,) and ecological assurance systems (including green buildings and environmental protection systems) are also necessary in this process.

Basic Characteristics A circular economy city is an urban system with the highest possible comprehensive benefits and a high conversion efficiency, which means that the living and material labor consumed in socioeconomic activities results in high economic achievements using the urban economic system with the lowest possible amount of resource inputs and waste generated. A circular economy is efficient and conserves resources and energy, producing minimal pollution and adhering to the 3 Rs. Specific characteristics of a circular economy include:

A Rational Industry Mix Led by Eco-friendly Industries The overall industry mix of a circular economy can be depicted as an inverted pyramid with the tertiary sector at the top and the primary at the bottom. Differences in natural conditions, historical backgrounds, and developmental orientations discourage requiring the same proportional relationships among these sectors in different cities. The tertiary sector should nevertheless occupy a certain minimum proportion because it reflects a city's vitality and the operating efficiency of its ecosystems. Some hold that the tertiary sector should occupy at least 70 % of the industry mix. All industries in a circular economy city should be green. The primary sector should focus on eco-agriculture, including green foods and ecological benefits. Large-scale agricultural industries that provide safe food and agritourism best accomplish this goal. The secondary sector should focus on eco-friendly industries. Eco-friendly industry chains, high-tech development, and resource and energy utilization at multiple levels best promote material conversion and regeneration. The tertiary sector should adopt green habits and provide a production environment conducive to eco-agriculture and eco-friendly industry development. Developing eco-friendly industries using a circular economy protects the city's environment and meets urban consumer demand.

An Efficient Economic Resource Circulation System A resource circulation system is at the core of the circular economy; it provides efficient transfer pathways for people, materials, energy, information, capital, and other economic resources using modern urban infrastructure to boost circulation and reduce economic resource loss and the resulting pollution of the urban environment. An efficient circulation system includes a three-dimensional transportation system with a subway system and elevated roads within the city as well as highways, aviation structures, railroads, marine options, and other transportation modes that connect the city with the rest of the world. It should also include a rapid information exchange system using digital communication, integration, and intelligence; a complete range of materials and energy supply systems, robust business, and financial service systems with reliable networks and rational distribution; and sewage and waste disposal along with other treatment systems equipped with advanced facilities. An eco-city using this blueprint will display circular economy values and ultimately provide a complete, healthy, closed-loop economic resource circulation system.

Conserving and Efficiently Using All Natural Resources and Energies Clean energy sources (such as solar, wind, and biogas) should be used to the maximum degree possible, and resource recycling should also be optimized to achieve cleaner production and green consumption. An eco-city should have a water resource recovery and reuse system. Development and use of biogas will save other energy sources and promote garbage and sewage recycling making it valuable to develop biogas applications where conditions allow. The number of high-rise buildings in cities provides highly feasible development of wind power generation. Wind power generation has certain advantages over biogas for eco-cities.

Lowest Possible Urban Facility Construction and Maintenance Costs Simple and practical building and decoration materials and technology should be favored over energy-intensive and flashy urban facility designs. Urban facilities serve people and require manual maintenance. Manual maintenance costs should be minimized and remain lower than the value of the services that the urban facilities provide. If a facility requires substantial human, material, or financial resources for construction or maintenance but can only provide rather limited services, it does not meet economic and ecological standards and falls far below circular economy standards.

6.2 Assessment Metrics of Circular Economy Cities

6.2.1 Principles

Assessment metrics for circular economy cities should be developed using the following principles:

Scientific Assessment metrics for circular economy cities should be based on eco-city development and objectively reflect circular economy principles. Each indicator's purpose, data collection methods, and calculation methods must be scientifically grounded. Indicators should be chosen using eco-city construction and circular economy theories.

Specific Assessment metrics for circular economy cities focus on circular economy city construction issues in a particular region and include indicators specific to that region that reflect regional characteristics and provide eco-city building guidance.

Systematic Indicators should cover economic, resource, environmental, social, and other factors. Assessment metrics for circular economy cities should look beyond economic factors to objectively and comprehensively describe general and particular subjects.

Hierarchy A circular economy urban system is complicated and large, requiring assessment metrics that reflect all aspects of circular economy urban development and that avoid indicator overlap. Indicators must be classified according to the urban system's structure to create an easy and accessible hierarchy.

Comparability Assessment metrics for circular economy cities should standardize and unify individual indicators on the basis of name, concept, equation, and statistical method in order to facilitate comparison.

Operability Indicators should be chosen with consideration of data availability, reliability, and measurability. Existing statistics and composite and key representative indicators will provide the best results. Assessment metrics for circular economy cities should balance simplicity and complexity, avoiding overly few or simple indicators that cannot adequately reflect a circular economy city and too many or overly complex indicators that hinder evaluation.

6.2.2 Developing Assessment Metrics for Circular Economy Cities

A circular economy city values the eco-friendliness of economic activities, protecting resources and the environment without jeopardizing economic development. Economic development, resource protection, environmental protection, and social progress must be factored into circular economy cities' assessment metrics because they will necessarily influence society. These principles and the eco-city indicator set were used to develop assessment metrics for circular economy cities (see Table 6.1).

6.3 Measures for Building Circular Economy Cities

Circular economy cities aim to develop within the natural ecosystem's carrying capacity; properly restrain demand and ensure effective supply; optimize natural resource allocation with the belief that "science and technology are the primary productive force"; transform the economic growth pattern using economic, legal, administrative, technological, and educational means, such as industrial restructuring, technological innovation, cleaner production, resource pricing, and green consumption; and combine efficient resource utilization and recycling with environmental protection to bring harmony to socioeconomic development and natural ecosystems, thus promoting the sustainable new economy.[1]

Circular economy cities offer a new urban development pattern that promotes socioeconomic development that effectively protects the natural environment and uses natural resources efficiently. They offer a key approach to organically combining industrialization, urbanization, and eco-friendliness. Urban economic activities must be integrated in the cycle leading resources to products and unto renewable resources to create positive interactions between the economic system and natural ecosystems. Economic activities should adhere to the 3 Rs in order to effectively use resources and protect the environment. These measures help eliminate the conflicts between urban economic development and environmental resources.

A circular economy generally involves circulation at three levels including:

Circulation at the Corporate Level Economic activities in a traditional economy push resources in an open system that develops resources, manufactures products, and finishes by generating waste, which leads to high resource consumption. Economic activities in a circular economy form a closed-loop circulation of resource that similarly develops resources, manufactures products, and finishes by generating waste but also by creating renewable resources, thus conserving resources.

Circulation at the Socioeconomic Level Producers (i.e., companies) invest capital, resources, technology, labor, etc., to make products demanded by consumers.

Table 6.1 Assessment metrics for circular economy cities

	Level 2 indicators	Unit	Standard value
Economic development	Per capita GDP	Yuan/person	33,000
	GDP growth	%	11
	Per capita revenue	Yuan/person	5,000
	Disposable income per urban resident per year	Yuan/person	16,000
	Net income per farmer per year	Yuan/person	8,000
	Percentage of the primary sector in GDP	%	10
	Percentage of the secondary sector in GDP	%	45
	Percentage of the tertiary sector in GDP	%	45
	Percentage of employees in the tertiary sector	%	40
	The percentage GDP used for new product and semi-industrial tests and subsidies for major research projects	%	1.5
Resource and environmental protection	Energy consumption per unit of GDP	Ton/1,000 yuan	0.14
	Water consumption per unit of GDP	Ton/1,000 yuan	15
	Pesticide consumption per unit area of farmland	kg/ha	4
	Chemical fertilizer consumption per unit area of farmland	kg/ha	250
	COD per unit of GDP	kg/ha	5
	SO_2 emission per unit of GDP	kg/ha	5
	The percentage of MSW converted into harmless substances	%	100
	The percentage of urban domestic sewage centrally treated	%	70
	The percentage of compliant industrial wastewater discharge	%	100
	The percentage of industrial solid waste that is recycled	%	80
	The percentage of industrial wastewater reused	%	50
Social progress	Engel coefficient	%	40
	Registered urban unemployment	%	4
	Urban air quality	Days compliant with level 2 standard or higher per year	200
	Per capita area of public green space	m^2	11
	Per capita road area	m^2	29

Waste generated by consumers are then processed by reproducers (i.e., natural ecosystems and eco-friendly companies) to once again become raw materials needed by producers for production. These are the basic steps of a circulatory production system in society.

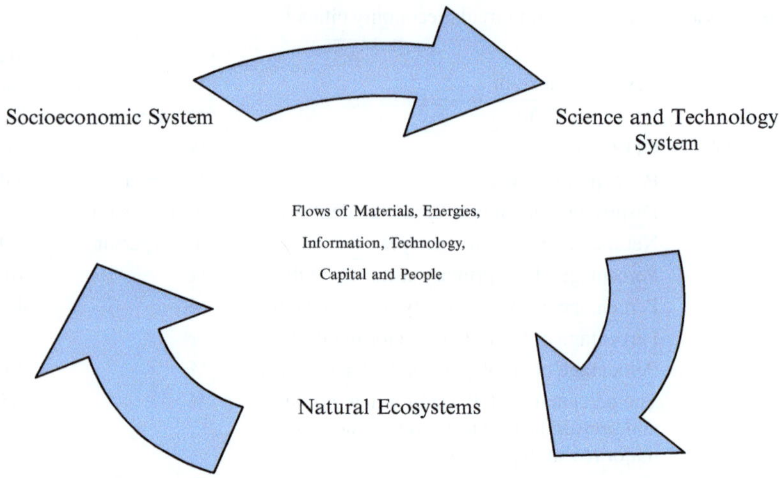

Fig. 6.1 Circulation through the socioeconomic system, science and technology system, and natural ecosystems

Figure 6.1 illustrates the circular flow of materials, energy, information, technology, capital, and people in the socioeconomic, scientific, and technological systems, as well as the natural ecosystems.

Circular economy development should be combined with eco-city development to create an entirely new, eco-friendly economic structure. New, eco-friendly industries are needed for this pattern of socioeconomic development. An eco-friendly economic system is needed, which includes the green agricultural and industrial sectors and eco-cities. The eco-friendliness of socioeconomic activities should be emphasized in urban planning.

6.3.1 Categories of Circulation in the City

Circulation in the city can be broken down into material and nonmaterial circulation, two separate categories that interact with each other. A number of measures can be taken to improve material circulation of products and throughout businesses, industries, and regions. Technology innovation and scientific design for circulation technology research can be enhanced. It is also important to manage waste as resources and to develop waste reuse strategies. Material consumption reduction per unit of output in production and economic activities is also critical for encouraging circulation. Finally, replace non-recyclable materials with recyclable alternatives. Nonmaterial circulation values orientation, knowledge innovation, information transmission, and industry chains to prime the environment for economic development, reduce resource consumption, and increase return on capital, thus providing support for material circulation.

6.3.2 Circulation Mechanism in the City

The laws and principles of circulation and natural ecosystems shape material and nonmaterial circulation chains.

The Eco-friendly Industry Chain The eco-friendly industry chain is a multichannel network of industrial connections between industries or companies that have a preexisting or potential correlation. It enables the flow of materials, information, and energy between industries and promotes interaction. The eco-friendly industry chain is built upon EIPs, such that advantageous resources and industries constitute the primary chain that other industries connect to in order to bolster their circulation.

The Policy Orientation Chain It is difficult for the eco-friendly industry chain to operate continually during the early development stages of a circular economy due to market price distortions caused by environmental externality costs. The government must develop economic and tax policies to orient the development of a circular economy and maximize socioeconomic and environmental benefits across society.

The Market Price Chain Circular economy development relies on the gradual formation of market prices dominated by mechanism promoted by the government. Industry ecosystems can be established in the early circular economy development stages using governmental power. Companies in a market economy will seek the highest value to maximize profits, and resources that can be put under the government's macro-control are also limited in a market economy. A corresponding market price chain control mechanism is necessary to promote a circular economy.

The Green Consumption Chain Green consumption and values are advocated in a circular economy to encourage the public to select environmentally friendly lifestyles and consumption patterns and to buy environmentally friendly products. This is an important task for developing a circular economy.

The Distribution Chain The distribution chain includes the information and logistics chains and is now in its earliest stage. Technical and information service systems as well as the information and logistic chains require development which can be achieved by establishing circular economy.

The Employment Chain A five-sector system is built in a circular economy by developing the environmental protection and waste recycling sectors in addition to the primary, secondary, and tertiary sectors. This system extends the employment chain and provides more jobs as well as diversifies services and human resource support for circular economy development and adding flexibility.

6.3.3 Circulation Participants in the City

Circulation participants in the city mainly include:

Resource Exploiters Circulation and circulation control begin with resource exploitation. Greater value will be created if resources are exploited in such a way that they can then be circulated.

Producers Companies act as producers and are best able to enact circulation. Their advantage comes from the fact that circulation relies on design and companies have the design capacity and the capability to decide which materials will be used for their products. Companies should assume responsibility for environmental pollution as the primary emitters and improve their technological capabilities to establish cleaner production, implement eco-friendly production processes, and reduce pollution. Resource recycling and environmental protection should be included in companies' innovation, development, and management strategies, and corresponding technical and managerial measures should be adopted in production, in management, and in other fields. Companies should be guided to best address a circular economy consumer and the corresponding market behavior.

Consumers Consumers can guide and discourage companies and their production activities since they select products and services. If consumers behave in an environmentally friendly manner and favor resource recycling, they push companies to implement circulation-based production. Completing material circulation in urban ecosystems also requires broad public participation. The public should establish eco-friendly values since it will benefit the most from circular economy benefits. People should accept green lifestyles and consumption patterns and cooperate with the government and businesses to build a circulation economy society.

Waste Treaters Organizations specializing in recovering, categorizing, treating, and reusing waste generated in the consumption process collectively compose waste treaters. They may have government or corporate backgrounds or function as independent organizations. Dual System Deutschland AG (DSD) is an example of an organization that is hired by companies to organize the recovery and categorization of their packaging wastes before sending them to resource recycling companies. Packaging wastes that can be directly reused are returned to manufacturers.

Service Providers These organizations contribute significantly to material circulation. The government should encourage individuals and financial, media, and other organizations to provide circulation services.

Scientists and Technologists These people play an important role in the knowledge economy by organically combining knowledge economy development with circular economy development and by establishing a green technical support system that is necessary for the eco-friendly industrial and agricultural sectors and the rest of society.

The Government The government can establish circular economy development as the basic strategic socioeconomic development objective. It should implement comprehensive plans and make appropriate laws, regulations, and policies to restrict behavior that does comport with a circular economy. The government should enact administrative, legal, economic, financial, and other means to advance macro-control efforts, maximizing the overall benefits of a circular economy.

6.4 Guiyang: A Circular Economy City

Guiyang is struggling with resource depletion, a low recycling rate, and fragile ecosystems like most areas in China. Guiyang decided to take the initiative and develop a circular economy to enhance sustainable development, rehabilitate the environment, significantly increase resource efficiency and people's incomes, and achieve harmony between humans and nature, making it China's first ever circular economy eco-city. Guiyang developed a circular economy plan to ensure it could achieve its goals. The basic circular economy development framework covers production and consumption pattern transformation, develops a circular economy industry system that incorporates the primary, secondary, and tertiary sectors; establishes urban infrastructure with a focus on water, energy, and solid waste recycling systems; and creates environmental support system including the living environment exemplified by green buildings and an environmental protection system.

6.4.1 Establishing a Circular Economy Industry System as Part of Circular Economy Eco-city Development

Eco-agriculture Efforts to build an eco-agricultural system in Guiyang focus primarily on developing eco-agricultural circulation and industrialization systems as well as locally unique eco-agriculture.

Building an Eco-agricultural Circulation System Guiyang has used recycling theories to blur the hard boundaries between the primary, secondary, and tertiary sectors in order to best use agricultural resources. The local government encourages rural residents to convert agricultural waste into raw materials used to produce biogas and provide a clean energy source. MSW is reused to form a closed-loop circulation system, thus improving the rural energy mix and the rural environment and increasing ecological and economic benefits.

Eco-agricultural Industrialization Eco-agricultural industrialization aims to replace rural household agricultural production with large-scale agricultural production that is more resource efficient. Eco-agricultural industrialization in Guiyang has advanced primarily in the following ways: Pollution-free agricultural products have been developed using contract farming. The local government encourages production and sale of pollution-free agricultural products that meet established standards. The city uses contract farming and agricultural technology demonstration zones to ensure a proportional development of farming, animal husbandry, and aquaculture, while developing pollution-free agriculture that satisfies people's need for healthy living. Pollution-free vegetable production centers are constructed. Guiyang boasts a large area of uncontaminated land that provides a unique natural environment for pollution-free vegetable production. Developing pollution-free vegetables can realistically industrialize vegetable production. Local organizations related to the pollution-free vegetable industry emphasize fresh varieties and varieties for processing purposes. These organizations have developed comprehensive vegetable processing and utilization mechanisms that focus on developing and applying technologies to better store and preserve vegetables, on producing vegetable juice and on providing precut, dehydrated, and frozen vegetables. These efforts will greatly improve vegetable resource utilization. Chinese herbal medicine resources are used to develop the traditional Chinese medicine industry. Guizhou Province is one of the main producers of traditional Chinese medicine. The local government promotes advanced extraction, purification, preparation, crystallization, and chiral technologies to develop high-tech products and accelerate the application of modern technology in traditional Chinese medicine production.

Developing Locally Unique Eco-agriculture Good agricultural practice (GAP)-based herb planting centers are constructed. Guiyang has aggressively promoted the standardization and industrialization of herb planting and processing and is developing a herb product quality management standard system to capitalize on local herb advantages. Local businesses have been applying advanced cultivation techniques and biotechnologies to improve the herb quality and yield. GAP-based herb cultivation centers are important in developing eco-agriculture and the processing industry because they contribute to economic development and environmental protection. Herb industry clusters follow environmental requirements tailored to specific varieties. The industry chain centers on farmers and industry clusters to achieve economies of scale. Eco-agricultural methods and technologies have been introduced and standardized production and management are used in planting, field management, harvesting, processing, follow-up, and waste treatment. Ecotourism and agritourism have been nurtured. The mountainous Guiyang found in Southwest China offers complex topography and ethnic customs and cultures both diverse and unique. Guiyang has great potential in developing tourism alongside its agricultural production. Guiyang focuses on developing three-dimensional agriculture that combines fish farming with lotus, vegetables, flat peach, and grapes, which helps with ecotourism and agritourism. Local organizations are building pollution-free, green food production centers at appropriate tourist spots that respect landscape ecology. Corresponding ancillary facilities are being built to attract people to the city for leisure and sightseeing. Guiyang is also striving to capitalize on agricultural activities typical of ethnic minority groups by leading tourists from around the world through an organic integration of ethnic customs, agritourism, and tourism resources. These efforts have contributed to Guiyang's reputation in pollution-free agricultural products and green foods.

An Eco-friendly Industrial Sector Local efforts to build an eco-friendly industrial sector are mostly present inside of enterprises, between enterprises, and at the city administration level.

Efforts in Enterprises Cleaner production and resource efficiency are promoted. A growing number of Guiyang-based industrial enterprises are using cleaner production methods. Local enterprises in traditional industries are accelerating technological improvement and increasing investment in technology innovation. The local government is imposing strict restrictions on small-scale enterprises with obsolete production processes and high energy consumption and pollution, effecting industries such as coal mining, papermaking, leather, cement, smelting, soda, calcium carbide, and yellow phosphorus. Measures have been taken to strictly control resource-consuming industrial projects. Phosphorus, aluminum, coal, and other mineral resource consumption, water demand, and the quantity of solid waste will be effectively reduced. Businesses are expanded and comprehensive by-product and waste utilization is increased. Many small Guiyang-based enterprises are unable to effectively utilize by-products and waste so that they remain unused or are discharged as pollution. By enlarging enterprises, the substances can be effectively used to more efficiently utilize resources.

Efforts Between Enterprises Correlated businesses are consolidated to improve overall economic benefits. The local phosphorus industry contains many enterprises responsible for phosphate mining and yellow phosphorus, fertilizer, and phosphate production. Despite a natural material stream between these industries, they lack proper consolidation. They possess great potential for improvement in economic efficiency, resource utilization, and environmental protection. Guiyang Gas Plant and Guizhou Crystal Organic Chemical (Group) Co., Ltd. are two coal chemical industry companies that are correlated with each other – the former produces coke and coke-oven gas and supplies them as energy sources and raw materials to the latter, which in turn provides the former with electric power and public works. These two companies have established a mutually beneficial relationship. Multiple industry partnerships, EIPs, and broad material and energy circulation will help enterprise coordination. EIPs are effective for comprehensively utilizing resources within the region, as compared with a single business or industry's industrial waste reuse capability. Cooperation among multiple industries further promotes industrial waste recycling. Partnerships between industries are also effective for optimizing resource use, promoting industry growth, and providing involved enterprises with positive sum benefits. Waste from yellow phosphorus production can be used to produce cement and can also be used to produce high value-added glass-ceramics processed at high temperatures. Tail gas from yellow phosphorus production can be used to produce dimethyl ether as a clean energy source, facilitating C1 chemical production, products that contain one carbon atom per molecule.

The City Administration Level Industrial and consumption system connections and a developed waste recycling sector are integral to a circular economy in Guiyang. The waste recycling sector enables lower energy and resource consumption in the production of raw materials. The energy needed to recycle aluminum is only 3 % of conventional aluminum production. Guiyang's large aluminum companies provide an opportunity to benefit the local economy and the companies themselves by using their equipment and technology advantages in recovering scrapped aluminum products and converting them to resources for all of Guizhou Province.

The city can advance circulation by participating in economic activities in Guizhou Province and in China. Guiyang industrial enterprises should expand beyond small closed circles in the city-administered areas to join economic activities at the provincial, national, and global level. Guiyang businesses produce ripple effects in business cooperation in the surrounding cities and prefectures, accelerating economic development in Guizhou Province. Economic activities should center on locally advantageous resources and energy sources as well as the city's position within China's economic strategy. Guiyang businesses should improve product quality, and the municipal government should offer incentives to grow competitive industries that can survive in the international market.

6.4.2 Increasing Infrastructure Construction for a Circular Economy Eco-city

The Water Circulation System A water circulation system is a primary symbol of a circular economy city. The water system is a circulation system of interactions between suppliers and consumers that efficiently allocates water resources at the lowest possible cost to reduce the impact of socioeconomic activities on water resources and the overall environment.

Building a water circulation system in Guiyang requires improving the water system and water infrastructure.

Building a Water System Guiyang's water system focuses on implementing water conservation projects in agriculture, daily life, and industry. The first water conservation project in agriculture was to change the irrigation method. The local government first identified about 1.5 ha of farmland in Xifeng County in 2003 for input–output and feasibility analysis of water-efficient irrigation technologies such as sprinkler, pipe, and drip irrigation. Water saving technology such as underground water tank sprinkler irrigation and drip irrigation was then gradually introduced into Xifeng, Qingzhen, Xiuwen, Kaiyang, and the suburbs of Guiyang to improve agricultural irrigation efficiency. Other efforts have included reducing leakage by replacing traditional ditches with channels lined with cement or other materials or with piping systems to reduce water lost due to seepage and evaporation, bringing loss down to 10 % or less. Measures such as ground leveling and loosening deep soil improve the soil's water storage capacity. Eco-friendly measures including horizontal ridging, soil covering, terracing, bio-fencing, etc. have further developed rainfed agriculture by conserving water and soil. Many efforts attempt to rehabilitate forest ecosystems since they are fundamental for environmental improvement.

Water conservation in daily life primarily consists of water saving equipment and device application. Apartment complexes, public toilets, and other public facilities built since 2003 are all equipped with water saving faucets, efficient flush tanks, and low-flow showers. 596 public toilets with flush tanks that were built before 2003 are gradually equipped with water-efficient flush tanks and slow-acting or infrared automatically controlled flush valves for urinals. All medium and large local hotels now use water saving faucets, flush tanks, and showers. Water meters have become important tools – water consumption in public buildings is now measured with water meters installed on incoming water pipes as is water consumption of home users.

Water conservation efforts in the industrial sector focus on electricity, chemistry, metallurgy, transport equipment manufacturing, beverage and food, and papermaking. Water conservation efforts in ethanol manufacturing before 2005 focused on promoting the dual enzyme method for starch fermentation and water saving cooling equipment, developing and applying high temperature yeast, promoting the bacterial fermentation process, and conserving cooling water. Water conservation efforts in the beer industry before 2005 focused on promoting high-concentration saccharification and fermentation technology and reducing the consumption of cooling water. Water conservation efforts in the canning industry before 2005 focused on promoting advanced water saving canning technology, the high back flow spiral cooling process, as well as water saving cleaning and sterilization processes. Recycling technologies are expected to increase the reuse rate in beverage manufacturing to over 80 % from the current 56 % and to over 70 % from 2.1 % in food manufacturing. Due to pollution control and restrictions on breweries with a capacity of 100,000 t per year, the COD removal rate is expected to increase to over 70 % from less than 35 % now and the percentage of compliant wastewater discharge is expected to increase to 100 % from the current 16 %.

Building Water Infrastructure Guiyang utility companies have been enhancing routine pipe inspection, repair, and replacement to prevent water loss and have improved water supply facility construction to further the circular economy and eco-city development. Resilient sewage treatment facilities including sewer networks and ancillary facilities are required to ensure sewage collection and transportation. Newly built sewer networks are required to carry rainwater and sewage separately, and old ones should be improved to achieve the same. Rehabilitating contaminated rivers and lakes should be prioritized, requiring sewage interception ditches.

The Solid Waste Recycling System The solid waste recycling system in Guiyang includes industrial solid waste reuse, MSW reuse, and solid waste disposal subsystems.

The Industrial Solid Waste Reuse Subsystem Large quantities of industrial waste cause a high environmental impact but have a high percentage of reusable component parts making recycling industrial waste an important means of improving resource utilization and reducing environmental pressures. Industrial solid waste in Guiyang includes phosphorus slag, phosphogypsum, red mud, fly ash, and coal gangue, which account for over 80 % of all industrial solid waste in Guiyang. Improving the comprehensive reutilization of these industrial wastes and controlling their generation is key to controlling the total amount of industrial solid waste. Solid waste should be temporarily stored before it is reused, requiring state provisions to prevent secondary pollution during storage. Red mud can be efficiently reused to produce new wall materials. Using red mud and fly ash to make sintered bricks is a simple process with reliable equipment and easy operation. The technology enables land conservation and environmental improvement because of the high consumption of red mud. Red mud production of new wall materials can also create positive economic benefits and provide a new source of economic growth.

The MSW Reuse Subsystem Building an MSW reuse subsystem in Guiyang includes:

MSW Categorization MSW in Guiyang is categorized into household waste generated in residential areas, business waste generated in business areas, office waste generated in education and research areas, and public waste generated in cultural areas, entertainment areas, and sports centers, as well as waste generated on roads and in parks.

Used Material Recycling Guiyang has established used material and renewable resource trading centers and a renewable resource utilization research center in an effort to standardize and improve the used material recovery network system alongside the MSW categorization. Local laws and regulations regarding used material recycling cover recycling of packaging containers, household appliances, and construction equipment, as well as the treatment of food resources. Used material trading centers are intended to standardize the used material marketplace. The renewable resource trading center consists of waste zones where scrap metal, waste paper, glass, plastics, and rubber are traded. The renewable resource utilization research center explores development and utilization of renewable resources, renewable resource information dissemination, and the promotion of renewable resource technologies.

The Solid Waste Disposal Subsystem This subsystem mainly includes hazardous waste disposal, medical waste incineration, and MSW disposal mechanisms.

6.4.3 The Living Environment, Green Buildings, and Environmental Protection in a Circular Economy Eco-city

The Living Environment Efforts to improve the living environment in Guiyang include building urban eco-friendly zones, planning land use in urban areas, improving the urban road network, and enhancing urban landscaping.

Green Buildings Guiyang, local authorities, and organizations focus on two issues related to green building projects – determining the relationship between buildings and their surroundings, as in how to minimize environmental damage caused by buildings, and identifying building materials and methods suited to Guiyang's characteristics. Local green building projects consider the hilly and mountainous terrain typical in Guiyang and use methods that enable land conservation while minimizing damages to natural resources.

Environmental Protection Environmental protection measures in Guiyang include creating environmental protection, buffer, and rehabilitation zones. Environmental protection zones require establishing water resource, natural ecosystem, and wildlife protection zones. Guiyang's water resource protection zones focus on environmental rehabilitation projects around the Hongfeng, Baihua, and Aha lakes, which are the three main sources of drinking water. Natural ecosystem protection zones focus on environmental protection and rehabilitation initiatives, such as protecting natural forests and preserving and cultivating forest resources. Environmental buffer zones require environment protection initiatives for the Nanming, Maotiao, Yuliang, Gucha, Yangshui, Qingshui, and Xifeng River basins, which are all part of the Yangtze River basin, and protection for the Mengjiang River basin, which is part of the Zhujiang River basin in the Guiyang-administered area. Local authorities and organizations run integrated rehabilitation programs covering mountains, water, land, forests, and roads to provide for ecological control zones. Biological and engineering measures accelerate afforestation. Efforts to preserve basic farmland are combined with afforestation to conserve water and soil, protect against winds, and stabilize sand. Local authorities have been working to establish economies of scale in the primary sector to increase farmer incomes. Action has been taken to reduce the pressures on land and water resources and to afforest barren hills and mountains as part of local efforts to form a structurally reasonable vegetation system.

Environmental rehabilitation zones involve urban and rural fringes and new built-up areas, such as the Jinyang New District. Plans for these areas adhere to high standards and concepts including zero-emission targets and the circular economy. Harmony between the natural and living environments and between natural and cultural landscapes is emphasized. All areas vulnerable to soil erosion and rocky desertification have been classified as forest and grassland protection areas to safeguard the integrity of urban green spaces and the functional stability and diversity of ecosystems. All farmlands with a slope greater than 25° have been included into afforestation projects. Bare or abandoned land and former quarry and mining sites have been planned as reclamation areas. Main water sources, rivers, and identified "no-build zones" have been designated for ecological rehabilitation to ensure water conservation and landscaping while preventing the functional degradation of ecosystems. Environmental rehabilitation zones have been identified along the Nanming River and south of the Aha Reservoir in Huaxi District, along the rivers in Longdongbao District, and in the northern part of Baiyun District. Residents in old urban areas and rural migrants are systematically relocated to new urban areas (Table 6.2).

Table 6.2 Top 50 Chinese cities by overall index of circular economy city, 2008–2011

2008		2009		2010		2011	
City	Ranking	City	Ranking	City	Ranking	City	Ranking
Shenzhen	1	Shanghai	1	Shanghai	1	Shenzhen	1
Shanghai	2	Shenzhen	2	Shenzhen	2	Shanghai	2
Nanjing	3	Nanjing	3	Nanjing	3	Hangzhou	3
Beijing	4	Beijing	4	Hangzhou	4	Nanjing	4
Suzhou	5	Hangzhou	5	Beijing	5	Beijing	5
Hangzhou	6	Suzhou	6	Suzhou	6	Chongqing	6
Yantai	7	Tianjin	7	Yantai	7	Suzhou	7
Jinan	8	Xiamen	8	Tianjin	8	Tianjin	8
Tianjin	9	Wuhan	9	Wuhan	9	Yantai	9
Wuxi	10	Zhuhai	10	Jinan	10	Wuhan	10
Wuhan	11	Jinan	11	Qingdao	11	Zhuhai	11
Xiamen	12	Wuxi	12	Zhuhai	12	Changsha	12
Zhuhai	13	Yantai	13	Shenyang	13	Jinan	13
Qingdao	14	Qingdao	14	Wuxi	14	Dalian	14
Chengdu	15	Dalian	15	Dalian	15	Xiamen	15
Ningbo	16	Baotou	16	Xiamen	16	Wuxi	16
Chongqing	17	Ningbo	17	Changsha	17	Nanchang	17
Shenyang	18	Chongqing	18	Changchun	18	Shenyang	18
Dalian	19	Nanning	19	Ningbo	19	Hefei	19
Benxi	20	Chengdu	20	Weihai	20	Ningbo	20
Changsha	21	Jiaxing	21	Hefei	21	Changchun	21
Changchun	22	Changsha	22	Harbin	22	Jiaxing	22
Jiaxing	23	Nanchang	23	Fuzhou	23	Chengdu	23
Nanning	24	Changchun	24	Nanning	24	Fuzhou	24
Weihai	25	Hefei	25	Jiaxing	25	Handan	25

Zhengzhou	26	Fuzhou	26	Handan	26	Shaoxing	26
Shaoxing	27	Shijiazhuang	27	Zhangjiajie	27	Xi'an	27
Baotou	28	Jining	28	Shaoxing	28	Jining	28
Zhangjiajie	29	Weihai	29	Chongqing	29	Weihai	29
Nanchang	30	Xuzhou	30	Nanchang	30	Shijiazhuang	30
Jining	31	Harbin	31	Jining	31	Quanzhou	31
Xuzhou	32	Zhengzhou	32	Baotou	32	Yinchuan	32
Fuzhou	33	Shenyang	33	Chengdu	33	Baotou	33
Harbin	34	Handan	34	Shijiazhuang	34	Harbin	34
Shijiazhuang	35	Baoding	35	Xi'an	35	Urumqi	35
Handan	36	Haikou	36	Quanzhou	36	Nanning	36
Hefei	37	Shaoxing	37	Zhengzhou	37	Kunming	37
Quanzhou	38	Qinhuangdao	38	Yinchuan	38	Qingdao	38
Baoding	39	Yinchuan	39	Qinhuangdao	39	Jilin	39
Haikou	40	Zhangjiajie	40	Xuzhou	40	Daqing	40
Ma'anshan	41	Ma'anshan	41	Haikou	41	Zhengzhou	41
Xi'an	42	Quanzhou	42	Kunming	42	Lianyungang	42
Yinchuan	43	Urumqi	43	Daqing	43	Karamay	43
Qinhuangdao	44	Huzhou	44	Huzhou	44	Haikou	44
Xiangtan	45	Xi'an	45	Lianyungang	45	Qinhuangdao	45
Huzhou	46	Kunming	46	Ma'anshan	46	Xuzhou	46
Lianyungang	47	Benxi	47	Guiyang	47	Huzhou	47
Daqing	48	Zhongshan	48	Baoding	48	Pingdingshan	48
Luoyang	49	Xiangtan	49	Jinchang	49	Baoding	49
Pingdingshan	50	Jiaozuo	50	Jiaozuo	50	Guiyang	50

References

1. Jiang Q (1995) Introduction to Gongyang School. Liaoning Education Press, Shenyang
2. Zhu D (2000) From sustainable development to circular economy. World Environ 3:6–12
3. Zhang Y et al (2005) Introduction to circular economy. Hunan People's Publishing House, Changsha
4. Xu Y (2005) Recycling economy: global trends and practices in China. People's Publishing House, Beijing

Chapter 7
Leisure Landscape Cities

Taichun Wang and Juheng Li

7.1 Concept and Meaning

Landscape refers to different ideas in different disciplines such as landscaping theories and practices. Ordinary people typically think of landscapes as scenery or garden imagery, such as urban green spaces that serve aesthetic and useful purposes, or natural landscapes. Landscape is defined both in a narrow and broad way in ecology. Landscape narrowly refers to heterogeneous geographic units that are composed of different ecosystems and exhibit repetitive patterns within a radius ranging from dozens to hundreds of kilometers. A landscape complex that reflects the climatic, geographic, biological, economic, social, and cultural features is referred to as a region. The narrow definition of a landscape and a region are usually referred to as macroscopic landscape. Landscape broadly includes microscopic and macroscopic spatial units that can be patchy or heterogeneous.[1] Urban landscape patterns and urban landscape element combinations must be viewed as an integrated landscape system to avoid one-dimensional thinking that ignores either the microscopic or macroscopic scale and provide for scientific and rational urban landscapes. The urban landscape system can thus be distinctive and fascinated and push the urban landscape pattern to prioritize people.

Leisure landscape results from advanced development of human civilization and takes the form of outdoor activities such as sightseeing, recreation, and travel in which people use and enjoy natural and cultural landscapes. Cultural activities in urban areas and nature reserves, lakes, beaches, and other tourist attractions in out

---

[1] Wu [1].

T. Wang (✉) • J. Li
Lanzhou City University, Lanzhou, China
e-mail: wangtaichun@126.com; 809050860@qq.com

J. Li, T. Yang (eds.), *China's Eco-city Construction*, Research Series on the Chinese Dream and China's Development Path,
DOI 10.1007/978-3-662-48153-0_7

suburbs and in mountainous areas provide a breadth of leisure landscapes. Maslow's hierarchy of needs proposes that people's needs can be represented as a pyramid with the more basic needs at the bottom, asserting that people will move to higher need levels after the most basic needs have been met. After people's monetary income increases to allow them to meet their basic life needs, they will seek higher levels of enjoyment such as leisure and recreation.[2]

Non-Chinese scholars have predicted that developed countries will enter the age of leisure around 2015 and that developing countries will closely follow. Leisure consumption will grow rapidly once per capita income reaches US $5000–8000.[3] China's per capita GDP reached US $8000 in 2010 and the national economy is growing steadily since recovering from the 2008 financial crisis. According to the National 12th Five-Year Plan, the per capita income of urban employees is expected to double and the level of consumption among the urban population will increase dramatically during that period. Leisure consumption in China is likely to enter a stage of rapid growth. Per capita income in many Chinese cities above the US $8000 level has laid the basic economic foundation for leisure consumption. The 2006 World Leisure Expo and the 9th World Leisure Conference were held in the Hangzhou World Leisure Expo Garden on October 15–20, 2006. "Let leisure change human life" was the slogan of Hangzhou World Leisure Expo. The theme "Leisure: an integral part of social, cultural and economic development" provided the background for the 9th World Leisure Conference, which was the highest-level international conference in the world's leisure and tourism industry, marking a new milestone in China's development of leisure cities. Leisure city refers to an advanced and modern city that has established a postmodern lifestyle. A leisure city is able to satisfy the general public's need for this lifestyle, exhibits postmodern cultural qualities, and forms an industry mix led by the leisure industry.[4] A clear global definition of leisure landscape city is not yet available, but many famous tourism cities in developed countries offer primarily urban leisure landscapes. Both locals and many domestic and foreign tourists enjoy the leisure lifestyle and sightseeing opportunities. Famous cities such as Paris, London, Venice, and Washington DC attract millions of visitors each year, thanks to unique urban landscape features, including historic landmarks in Paris, water landscapes in Venice, historic and cultural landscapes in Rome, famous modern landscapes in New York and Dubai, natural landscapes in Hawaii, and large-scale, integrated, amusement landscapes exemplified by Disneyland Resorts in the United States, Japan, Hong Kong, and other countries and regions. A leisure landscape city is an eco-friendly city where urban landscapes are the primary sites for mostly outdoor leisure activities. This title is evidence of a city's overall strength and of its development in urban aesthetics, culture, and economy.

[2] Sun [2].

[3] Waldheim [3].

[4] Yang [4].

7.2 Developing Assessment Metrics for Leisure Landscape Cities

The environmental sustainability of a city relies on coordinated development of urban economic, social, and natural environments. A sustainable economic environment is a prerequisite of the overall system to develop sustainably. A sustainable natural environment ensures that the overall system is able to develop sustainably. A sustainable social environment is the goal of the overall system's sustainable development.[5] Green spaces in China are currently assessed and planned using only three quantitative indicators: per capita area of public green space, urban vegetation coverage, and urban green space rate. The limited information does not allow for a full evaluation of the landscape and of social and economic benefits of urban green spaces. Quantities have long been prioritized over benefits in comprehensive evaluation of urban green spaces.[6] These assessment metrics cannot reflect the quality of urban landscapes, the scientific rationality behind pattern and structure, or the ways they serve people.

7.2.1 *Principles*

Assessment metrics for leisure landscape cities should follow these principles:

1. Indicators from social and natural sciences should complement each other (i.e., combine social surveys with mathematical calculation) to form scientifically rational assessment metrics.[7]
2. Indicators should reflect the city's landscape pattern by applying ecological principles in urban landscapes to reflect the pattern, structure, and ecological functions of urban landscapes as well as urban landscape qualities, including the quantity of landscape facilities, gardening culture, art, and the quality of landscaping services.
3. Indicators should be comprehensive and fully consider qualitative and quantitative operability and comparability.
4. Indicators should reflect the environmental quality of the city by including air and water quality and noise level.
5. Indicators should include integration of socioeconomic, landscape, and environmental factors for comprehensive assessment.
6. Indicators should provide a static and dynamic assessment model to present the past, present, and future of urban landscapes.

[5] Wang [5].

[6] Ryan [6].

[7] Rong et al. [7].

7.2.2 Assessment Metrics for Leisure Landscape Cities

Complete assessment metrics were developed using these principles including socioeconomic, environmental, and landscape measures as Level 1 composite indicators and 20 Level 2 indicators such as landscape diversity (see Table 7.1). These indicators are intended to truly reflect the city's essential characteristics and status in terms of socioeconomic and environmental development and in terms of leisure landscapes and to guide the city in developing a leisure landscape orientation.

N1–N9 in this table are Level 2 indicators under urban landscape and leisure; N10–N14, under urban environment; and N15–N20, under socioeconomic development.

7.3 Building Leisure Landscape Cities

7.3.1 Status Quo and Challenges

7.3.1.1 The Status Quo

Unclear Understanding of Urban Landscape Design Sustainable development must be able to satisfy the needs of the current generation and future generations, and it must be coordinated such that socioeconomic development considers environmental resources and carrying capacity. Applying this concept of sustainable development to urban landscape planning and design results in models that respect nature, people, and culture.[8] Chinese government departments and organizations researching and implementing modern urban landscape design and planning began this work only two decades ago, which has not been enough time to grant a comprehensive understanding of landscape design theories. City managers and landscape researchers and designers need further familiarization with urban landscape design theories and concepts.

Wide Variation in Urban Landscape Design Quality Urban landscape design involves urban architectural design (including building height, style, color, etc.), landscape design (including hard and soft landscapes), ecological design (including landscape ecology, environmental protection, botany, ecosystem science, community ecology, etc.), urban cultural design (including local culture, history, regional culture, folk culture, ethnic culture, etc.), and many other disciplines. The breadth requires an urban landscape design team to have strong cooperation capabilities and integrated research and design. Current urban landscape design teams in China vary greatly in their competence. Most of these teams are unable to systematically integrate landscape resources, aesthetic expressions, cultural expressions, human

[8] Wang [8].

Table 7.1 Assessment metrics for leisure landscape cities

Objective layer	Criterion layer	Indicator layer	Standard value	Reference	Level (in China): excellent, good, average, poor	Calculation and acquisition methods
		Landscape diversity (N1)	1	Maximum	>0.8; 0.7-0.8; 0.6-0.69; <0.6	$H' = 1 - \sum_{k=1}^{n} PK^2$ City satellite map
		Landscape patch connectivity (N2)	1	Maximum	>0.8; 0.7-0.8; 0.6-0.69; <0.6	$C = \left[\frac{\sum_{j-k}^{n} c_{jk}}{\frac{n_i(n_i-1)}{2}}\right] \times 100$
	Urban landscape and leisure (M1)	Landscape accessibility (N3)	95 %	Maximum	95 %; 85 %; 75 %; <65 %	The minimum cumulative resistance (MCR) model; ArcGIS spatial analysis; satellite map
		Landscape distribution and satisfaction with facilities/services (N5)		The mean of questionnaires	Excellent, good, average, poor	Refer to the Urban Beauty Questionnaire
		The number of theaters per 100,000 people (N6)	1	Temporary	>1; 0.8-1; 0.6-0.79; <0.6	*China City Statistical Yearbook*
		The number of sport venues per 100,000 people (N7)	1.2	Temporary	>1.2; 0.9-1.2; 0.7-0.89; <0.7	*China City Statistical Yearbook*

(continued)

Table 7.1 (continued)

Objective layer	Criterion layer	Indicator layer	Standard value	Reference	Level (in China): excellent, good, average, poor	Calculation and acquisition methods
		The number of buses per 10,000 people (N8)	5	Temporary	>5; 4-5; 3-3.9; <3	*China City Statistical Yearbook*
		The urban green space rate (N9)	45 m^2	Global cities	38; 30; 25; <25 m^2	(The total urban green space area/the total area of built-up land) × 100 %
Leisure landscape city		The urban forest coverage (N10)	50 %	Global cities	>40; 30; 25; <20	[(The total area covered by woodlands including commercial woods, bamboo forests, and shrub lands each with a canopy density of at least 0.3, farmland shelter belts and trees beside villages, houses, bodies of water, and roads)/the urban land area] × 100 %
	Urban environment (M2)	The urban air indicator (N12) (Grade II)	300d	China's garden city standard	Excellent (300d); good (270d); average (250d); poor (<230d)	The urban environment section of China's garden city standard
		The percentage of urban sewage treated (N13)	90 %	Temporary	>90 %; 85-90 %; 75-84 %; <75 %	*China City Statistical Yearbook*
		The ambient noise compliance rate (N14)	95 %	China's garden city standard	95 %; 85 %; 80 %; <70 %	The report on the urban environment

		The urban GDP per capita (N15)	US $3000	Temporary	>4500; 3000-4500; 2000-2999; <2000	*China City Statistical Yearbook*
		The GDP growth rate (N16)	9 %	Temporary	>15 %; 10-14.9 %; 8-9.9 %; <8 %	*China City Statistical Yearbook*
	Socioeconomic development (M3)	The per capita road area (N17)	30 m^2	Temporary	>30 m^2; 25-30 m^2; 15-24.9 m^2; <15 m^2	*China City Statistical Yearbook*
		The percentage of the tertiary sector in GDP (N19)	45	Temporary	<30; 31-40; 41-45	*China City Statistical Yearbook*
		The Engel coefficient (N20)	30 % means the richest	International standard	Poor (>59 %); subsistence (50-59 %); well-off (40-50 %); rich (<40 %)	*China City Statistical Yearbook*

considerations, and ecosystem services. Urban landscape design teams in China must become more competent to provide designs with sufficient quality.

Wide Variation in Urban Landscape Quality Economically developed cities tend to have better urban landscapes than economically underdeveloped cities due to differing design ideas and investment. Landscape quality variation affects selected materials and workmanship. Economic factors as well as varying local government management styles and landscape design philosophies lead to these variations. Urban landscaping involves a long-term plan implemented step by step. Haste and comparison lead to an unexpected lack of harmony between urban landscapes and the overall urban spatial structure.

Urban Landscapes with no Outstanding Features Urban landscapes are the face of a city. They comprehensively reflect the city's geographical and topographical qualities as well as its water resources, buildings, plant population, history, culture, ethnicity, etc. Every city has unique landscape resources; historical, regional, and cultural characteristics; and development process. Discussions among all relevant parties and careful study are required for the admirable goal of building a distinctive city using its unique characteristics. City managers should address the fact that many Chinese cities look alike using scientific planning.

7.3.1.2 Challenges for Building Leisure Landscape Cities in China

The Conflict Between Urban Development and Landscaping Urban expansion and development require a lot of land. The housing market, industry clusters, and commercial buildings provide cities with huge economic benefits. Spaces originally reserved for urban landscaping are often used for economic benefits, ignoring the importance of parks and public green spaces in the urban landscape system. City managers are attempting to resolve the growing land conflict between landscaping and construction under the Scientific Outlook on Development.

Urban Ecosystem Protection is Urgent Urban ecosystem protection involves ecological protection zones along the city periphery, in ecologically sensitive areas, and on suburban farmland. Urban ecosystems improve citizens' living environment by cleaning the air and water, reducing noise, adjusting temperatures, and preventing soil erosion. Urban landscapes constitute an important part of urban ecosystems. Many Chinese cities struggle with a deteriorating urban environment due to resource overuse and massive industrial waste gas and water emissions. Improving urban ecosystems and urban landscapes is a key component of urban development, safety, and human health.

Rebuilding Old Urban Areas and Improving the Urban Landscape System A city develops dynamically. Old urban areas may need new planning and construction over time. City managers need to consider how urban landscaping and restoration of old urban areas can be combined to provide the city with a more appropriate spatial structure, restore ecological functions, and improve citizens' living environment.

7.3.1.3 Tasks for a Leisure Landscape City

Scientific Planning and Design for the Urban Landscape Pattern Urban economic growth fueled by building locally distinctive landscapes to attract more tourists to the city provides many urban development benefits.

Protecting and Utilizing Landscape Resources and the Local History and Culture Urban culture is both modern and traditional. Both aspects must be combined by preserving the traditional culture and developing modern culture. The city's qualities and its desired future should be represented in the form and language of landscape culture.[9] Preserving old streets, residential buildings, temples, estates, and other buildings that represent various developmental stages and major historic events of the city is key to sustaining traditional culture. Tourist attractions including natural landscape resources such as old trees, rocks, and springs that are connected to local legends should also be cared for. A series of city landscape spaces and a landscape culture should use the cultural elements reflected in traditional buildings and the characteristics of local natural landscape resources to give itself form. Preserved urban landscapes should be connected by roads into a network, making the urban landscape pattern more accessible.

Building Landscape Service System for People Improving greenways, teahouses, parking lots, trash cans, green space features, benches and chairs, lighting systems, and other outdoor amenities ensures landscapes best serve the public. The relationship between landscape and residential area distribution must be considered when designing landscape patterns so that the public can enjoy a leisure landscape lifestyle near their homes.

Creating a Landscape Culture System Every city has its particular development process and inherent characteristics based on regional resources and a specific urban culture. Examples include the Silk Road, the Yellow River, the Yangtze River, democratic, folk, and architectural culture. Identifying and combining these cultural elements with local natural landscape resources allows one to visualize the city's landscape culture system in an artistic landscape design language.

Advocating for a Leisure Landscape Lifestyle and Developing the Leisure Landscape Industry in the City The municipal government should provide policy support for the leisure landscape industry in the form of tax breaks and space allotment and should mobilize media resources to publicize leisure landscape to instill it in urban residents' consumption. This will boost the industry to provide major support for urban economic growth.

[9] Xiong [9].

7.3.1.4 Characteristics and Varieties of Leisure Landscape Cities in China

Some Chinese cities are known for their history and culture, such as Beijing and Xi'an. In the city of Beijing, the Forbidden City, the Great Wall, the Summer Palace, the Temple of Heaven, the Old Summer Palace, and other historic sites and cultural landscapes constitute a unique urban landscape system. These historic and cultural landscapes are Beijing's most attractive features, beating modern parks and landscapes in popularity. In the city of Xi'an, the ancient city wall, the Big and Small Wild Goose Pagodas, the Bell and Drum Towers, the Terracotta Warriors, the Qin Shi Huang Mausoleum, the Huaqing Hot Springs, and a series of legends make up the soul of local urban landscapes. These landscapes contain the respective histories and cultures of the two cities, attracting domestic and foreign tourists.

Some Chinese cities are known for natural landscapes, such as Guilin. Natural landscapes experience only indirect, slight, or occasional human interaction and remain largely intact. Examples include the north and south poles, mountains, large deserts and swamps, tropical rainforests, and some nature reserves. Returning to nature is a growing force in the tourism and leisure industries. The saying "east or west, Guilin landscape is best" demonstrates Guilin's reputation for natural landscapes in China. Its beautiful natural landscapes present visitors with picturesque creations of God.

Some Chinese cities are known for cultural landscapes, such as Suzhou and Hangzhou. Cultural landscapes are built upon natural landscapes but have literary, artistic, and other cultural elements added to meet people's material and spiritual needs. Examples of cultural landscapes include famous attractions such as the Lingering, Lion Forest, Humble Administrator's Gardens, and the Tiger Hill in Suzhou as well as the West Lake in Hangzhou. These landscapes provide a glimpse of former royal and private gardens in China and demonstrate the essence of classical Chinese gardens. These landscapes follow the idea "from and beyond nature," integrating with the poetic and pictorial elements of the Chinese culture, making them unique in the world history of landscaping. They further offer cultural symbols for the city, providing a unique view of its landscape culture.

Some Chinese cities are known for modern urban landscape features, such as Shanghai and Shenzhen. Shanghai has been a bustling metropolis for over a hundred years, with international trade, finance, and shipping at the core of the city's industry mix. Major local urban landscapes include famous skyscrapers such as the Oriental Pearl and the Shanghai World Financial Center, busy shopping streets, the Huangpu River waterfront, and other landscapes that contribute to Shanghai's image as one of the world's most important cities. Shenzhen has emerged as an industrial city since China began reform and opening up and has come to represent China's modern cities. Shenzhen has a short history as a city but epitomizes China's achievements since reform and opening up. Shenzhen boasts landscapes typical of a bustling city that emulate high-tech development and modernization.

Other Chinese cities are known for ethnic landscapes, such as Lhasa and Dali. Multiple ethnic groups make their home in China. Various ethnic architectural styles, cultures, and customs form the core of ethnic minority landscapes and provide unique attributes to landscapes in cities such as Lhasa and Dali. Visitors to Dali are intoxicated with locally unique buildings, costumes, folk performances, and natural scenery. Visitors to Lhasa similarly admire the blue skies and white clouds typical on the plateau as well as the Tibetan character, costumes, houses, and customs. The combination of ethnic customs, architectural styles, cultures, and landscapes permits these cities to present a distinctive picture of urban landscapes.

7.3.2 *Measures for Building Leisure Landscape Cities*

7.3.2.1 Problems and Challenges

Lack of Scientific and Appropriate Urban Landscape Design Urban development takes a long time. Common urban problems such as traffic congestion, unfitting industry mix, severe pollution, and charmless landscapes stem from a lack of long-term, sustainable urban planning. Urban landscape pattern planning is particularly marginalized in the pursuit of economic growth. Long-term planning for a complex urban system requires close teamwork among experts from multiple disciplines and cannot be accomplished with short-term actions. A long-term urban plan should be thoroughly implemented step by step, and governmental changes should avoid interfering with the plan implementation.

The City's Landscape Culture and Resources Are Improperly Combined Tall buildings have appeared in dense groupings in many Chinese cities in recent years. The city's landscape pattern and culture are often ignored as this type of development expands. An inappropriate urban landscape plan or the lack of a plan disconnects urban landscapes and the local culture, damages the integrity of the landscape pattern, and leads to the disappearance of landscape heritage. Some city managers believe landscapes are restricted to parks and green spaces, ignoring the urban landscape culture system. They are anxious to conduct industrial, business, and commercial real estate projects and neglect urban landscape and landscape culture system design. Some landscape resources such as traditional houses, old streets, and locally distinctive cultural landscapes have been damaged or destroyed in this process. Tourist attractions that carry a historic significance for the city end up surrounded by incompatible modern buildings. These phenomena have become common in many cities in recent years. Urban landscapes should reflect the history of the city and its unique regional, cultural, and geographic features, telling people about the city's cultural context and development process. The urban landscape and landscape culture systems should never be overlooked to favor economic development and rapid urban expansion because these systems are an integral part of the city's culture and history.

Landscaping Budget Financial difficulties pose a serious problem in landscaping, especially in economically underdeveloped regions. Many cities faced unemployment, economic development difficulties, obstacles to urban renewal, and many other challenges, leaving them insufficient fund for investment in developing the urban landscape and landscape culture systems. City managers should value the rational protection, development, and use of landscape resources as well as the long-term planning and design of urban landscapes when thinking about urban development and management despite other concerns. They should properly fund phased implementation of long-term plans and avoid white elephant projects such as city squares and landscape avenues that are incompatible with the overall landscape culture and humanity.

Policy Guidance and Urban Management Philosophy Many Chinese cities have long held economic growth as the primary goal for urban development, especially since China began the reform and opening up. The long-term urban industrial plan and the urban development pattern, including urban landscapes and the leisure landscape industry, are often unscientific or left incomplete despite reflecting important city characteristics. Urban landscapes go beyond city image to present a larger picture of various urban elements such as ecological safety, rational allocation of resources, urban pattern, urban environment, and sustainable urban development. A scientifically rational urban landscape plan and design that shows the city's characteristics reflects wisdom and maturity in the management team. Urban development relies on urban expansion but must also focus on establishing a good urban landscape culture.

7.3.2.2 Measures for Building Leisure Landscape Cities

Enhancing Theoretical Research on Leisure Landscape City Construction Urban landscape research in China focuses on two issues. The first issue is the macroscopic urban ecological pattern thought about from a landscape ecology perspective. Research on this issue falls into three basic categories: the landscape structure, including the type and diversity of landscape units and the spatial relations among them; landscape functionality, namely, the interactions between the landscape structure and ecological processes or interactions among landscape units; and landscape dynamics, which refers to structural and functional changes in landscapes over time. The city is divided into patches, corridors (i.e., roads and rivers), and the substrate (i.e., groups of buildings in large areas of the city and hard surfaces) for the sake of large-scale research. Research results provide general guidance for urban landscaping. The other issue is the planning and design of small and medium landscapes. Research on this issue covers roads, waterfront landscapes, residential areas, parks, squares, public green spaces, and historically and culturally significant sites. Research efforts focus on the cultural, artistic, and practical properties of landscape space series as combinations of landscape elements such as terrains, ground, plants, bodies of water, landscape buildings, green space features, and public facilities. The first issue is researched by ecologists and the second by landscape designers. Urban landscaping is a comprehensive project that includes macroscopic to microscopic elements and urban landscape research efforts at both levels should collaborate and avoid being led separately to ensure a complete picture of the city's landscapes. Conceptualizing leisure landscape cities provides a new research approach for determining how to design and build an integrated landscape system that is structurally appropriate, culturally and artistically appealing, and able to meet the public's recreation needs. Contributing to scientific and rational design layout and improving the urban landscape construction level is conducive to improving urban landscape designs.

Implementing Strategies and Policies for Leisure Landscape City Development An urban development strategy seeks to balance economic growth, ecological balance, and environmental development. The municipal government should develop a long-term plan and implement it step by step. The plan should be codified as is common practice in developed countries. This ensures that the plan is adhered to regardless of government leaders, which protects sustainable urban development from white elephant projects. Urban landscaping and leisurely lifestyles will become increasingly important to urban development as social progress and economic growth continue, requiring that they be incorporated in the government's urban development policy. Government authorities should plan urban landscaping as early as possible using the city's environmental resource characteristics to meet consumer needs for leisure landscapes. A leisure landscape industry development strategy is one expected addition to urban development.

Preserving the City's Historic and Cultural Resources While Developing Urban Landscape Resources China is known for its diverse terrains and long history. Each Chinese city has unique features, resources, and development processes. The development process leads to diverse urban history, culture, and landscapes. One of the most important modern urban development principles is to respect cultures, including historical, folk, democratic, and regional cultures. The Forbidden City in Beijing, the West Lake in Hangzhou, and the ancient city walls in Xi'an are examples of iconic urban symbols renowned globally. Since both history and culture are created by human beings, emerging cities can build unique urban landscapes as long as they protect and utilize the local cultural and landscape resources and implement appropriate planning and design. Qingdao, Dalian, and Shenzhen have been attracting visitors from around the world with their distinctive landscapes since China began the reform and opening up.

Creating an Environment for Desirable Consumption and Customs Customs are also an important factor of leisure landscape consumption. Visitors to Sichuan Province are impressed by ubiquitous shade under trees, teahouses, and recreational activities. Desirable customs and culture are fostered by bolstering material support for landscapes, such as landscape facilities, public green spaces, parks, and gardens. The municipal government should provide incentives for the outdoor leisure landscape industry, such as tax breaks and land allotment, to guide the industry's growth and gradually establish the desirable culture and habits.

7.4 Guangzhou: A Leisure Landscape City

China's rapidly growing society has embraced leisure landscape cities and what benefits they provide and planners have established assessment metrics to determine progress. A leisure landscape city is an advanced Chinese urban manifestation suited to the eco-civilization that has not been comprehensively studied anywhere in the world. No leisure landscape city identification and assessment systems currently exist in China. But leisure landscape cities will emerge in large numbers as eco-friendly cities as a result of ecological advancements and urban diversification. China boasts a long history of landscape-oriented architecture and landscape design and creation as well as abundant natural and cultural landscape resources. Distinctive urban landscape patterns have already appeared in China during urban development. The new era of eco-civilization can reasonably be expected to usher in leisure landscape cities as a powerful engine for urban diversification. Guangzhou is more distinctive and successful than most Chinese cities that have been working toward leisure landscape urban development. Guangzhou's experience and practices are detailed below. Greenways appeared in urban landscape design in the 1970s. The "green" refers to natural elements, such as forests, riverbanks, and wildlife and the "way" refers to corridors. Greenways are hence natural corridors that interact with artificial landscapes.

Greenways are green open spaces with limited access, usually built along the riverfront, in valleys, along ridges, beside scenic roads, next to railways, along canals, and in connection with other natural and artificial corridors, offering sightseeing routes for visitors and cyclists that connect with major highways, nature reserves, resorts, historic sites, and urban and rural residential areas. Greenways can be walking trails and paths for bicycles and other nonmotorized vehicles and include parking lots, car rental shops, tourist shops, snack bars, and other facilities, buffered with greenery. The 2010 Asian Games held in Shenzhen offered Guangzhou the opportunity to complete a fully equipped network of greenways in the Pearl River Delta with a total length of 2372 km on January 5, 2011. This network has been connected to local greenways in Dongguan, Foshan, and Zhongshan. Wang Yang, the then head of the Guangdong Provincial CPC Committee, proposed building greenways for ecological, environmental, public, and economic benefits and gave a 3-year timeline for the project. Greenways in Guangzhou cover 12 districts and counties, including 234 tourist attractions, 98 towns, 42 Asian Games venues, and 52 subway stations, making it the network with the longest routes, the widest coverage, the largest number of accessible attractions, the largest population served, the most comprehensive facilities and equipment, and the widest distribution in central urban areas out of all cities in the Pearl River Delta. Commitment to government-led public participation prompted Guangzhou's municipal government to allocate administrative responsibilities covering maintenance, sanitation, security, etc. to various departments and governmental levels. Businesses are responsible for commercial operations. The local network of greenways was phased into operation. This allowed for establishment of reliable operation and management mechanisms to

Table 7.2 Top 50 leisure landscape cities by overall index, 2008–2011

2008		2009		2010		2011	
City	Ranking	City	Ranking	City	Ranking	City	Ranking
Shenzhen	1	Shenzhen	1	Shenzhen	1	Shenzhen	1
Guangzhou	2	Guangzhou	2	Guangzhou	2	Xiamen	2
Nanjing	3	Shanghai	3	Shanghai	3	Beijing	3
Shanghai	4	Beijing	4	Beijing	4	Zhuhai	4
Beijing	5	Nanjing	5	Nanjing	5	Guangzhou	5
Xiamen	6	Xiamen	6	Zhuhai	6	Hangzhou	6
Zhuhai	7	Zhuhai	7	Hangzhou	7	Shanghai	7
Hangzhou	8	Hangzhou	8	Xiamen	8	Nanjing	8
Dalian	9	Dalian	9	Dalian	9	Sanya	9
Shenyang	10	Nanchang	10	Shenyang	10	Xi'an	10
Wuhan	11	Wuhan	11	Tianjin	11	Wuxi	11
Tianjin	12	Suzhou	12	Changchun	12	Wuhan	12
Suzhou	13	Changsha	13	Changsha	13	Qingdao	13
Wuxi	14	Tianjin	14	Nanchang	14	Hefei	14
Nanchang	15	Wuxi	15	Wuhan	15	Zhenjiang	15
Qingdao	16	Sanya	16	Sanya	16	Shenyang	16
Changchun	17	Qingdao	17	Qingdao	17	Nanchang	17
Changsha	18	Weihai	18	Hefei	18	Changsha	18
Chengdu	19	Baotou	19	Weihai	19	Fuzhou	19
Zhengzhou	20	Jiaxing	20	Suzhou	20	Changchun	20
Huangshan	21	Yantai	21	Wuxi	21	Guiyang	21
Yantai	22	Changchun	22	Xi'an	22	Baotou	22
Ningbo	23	Hefei	23	Zhenjiang	23	Huangshan	23
Fuzhou	24	Fuzhou	24	Yantai	24	Yantai	24
Hefei	25	Zhenjiang	25	Yinchuan	25	Yinchuan	25

(continued)

Table 7.2 (continued)

2008		2009		2010		2011	
City	Ranking	City	Ranking	City	Ranking	City	Ranking
Weihai	26	Chengdu	26	Fuzhou	26	Ningbo	26
Zhenjiang	27	Yinchuan	27	Jiaxing	27	Suzhou	27
Baotou	28	Huangshan	28	Baotou	28	Dalian	28
Jiaxing	29	Ningbo	29	Chengdu	29	Chengdu	29
Yinchuan	30	Shenyang	30	Huangshan	30	Qinhuangdao	30
Xi'an	31	Shaoxing	31	Ningbo	31	Weihai	31
Sanya	32	Zhengzhou	32	Zhengzhou	32	Jiaxing	32
Chongqing	33	Shijiazhuang	33	Shaoxing	33	Tianjin	33
Shaoxing	34	Qinhuangdao	34	Guiyang	34	Shijiazhuang	34
Qinhuangdao	35	Xi'an	35	Shijiazhuang	35	Kunming	35
Shijiazhuang	36	Guilin	36	Qinhuangdao	36	Shaoxing	36
Guilin	37	Chongqing	37	Chongqing	37	Lijiang	37
Guiyang	38	Meizhou	38	Guilin	38	Yichun, Jiangxi	38
Yichun, Jiangxi	39	Yingtan	39	Yingtan	39	Zhengzhou	39
Meizhou	40	Guiyang	40	Yichun, Jiangxi	40	Chongqing	40
Hanzhong	41	Yichun, Jiangxi	41	Meizhou	41	Mianyang	41
Lijiang	42	Mianyang	42	Kunming	42	Baoji	42
Yingtan	43	Kunming	43	Lijiang	43	Yueyang	43
Luoyang	44	Lijiang	44	Mianyang	44	Guilin	44
Mianyang	45	Yueyang	45	Yueyang	45	Meizhou	45
Yueyang	46	Hanzhong	46	Changzhi	46	Yingtan	46
Kunming	47	Changzhi	47	Hanzhong	47	Luoyang	47
Changzhi	48	Luoyang	48	Luoyang	48	Hanzhong	48
Baoji	49	Yichun, Heilongjiang	49	Baoji	49	Yichun, Heilongjiang	49
Yichun	50	Baoji	50	Yichun, Heilongjiang	50	Changzhi	50

ensure the greenways' sustainability. Parking lots, rural attractions, landscape nodes, and amusement facilities have been built along the greenways to provide accommodation, transportation, travel options, shopping opportunities, and amusement, thus meeting various needs of tourists and attracting local residents to the greenways. Greenways provide an unprecedented opportunity for promoting urban-rural integration, the growth of related industries, infrastructure expansion, public service provision, and urban civilization growth in rural areas. They have attracted many urban dwellers into rural areas for recreation, health, and tourism, fueling socioeconomic development in rural areas as well as urban-rural interactions and integration.

The greenway project has become an integrated urban landscape resource that is connected to a full range of supporting facilities. Eleven service centers have begun operating in conjunction with the project. Citizens have access to bicycles at these service centers and can return them at various locations across the city. These service centers increase connectivity and contribute to urban sightseeing and leisure consumption. Guangzhou's experience from the greenway project offers an approach to building a leisure landscape city (Table 7.2).

References

1. Wu J (2007) Landscape ecology: pattern, process, scale, and hierarchy. Higher Education Press, Beijing
2. Sun T (2003) On Yantai's advantages and strategies in building a leisure-oriented city. Urban Probl (6)
3. Waldheim C (2006) A reference manifesto. In: Waldheim C (ed) The landscape urbanism reader. Princeton Architectural Press, New York, pp 11–19
4. Yang S (2000) On the environmental sustainability of cities., J South China Norm Univ (Nat Sci Ed) (4)
5. Wang R (1988) Efficiency and harmony: the principles and methods of urban ecological regulation. Hunan Education Publishing House, Changsha, pp 268–269
6. Ryan RL (2011) The social landscape of planning: integrating social and perceptual research with spatial planning information. Landsc Urban Plan 100:361–363
7. Rong B, Chen C, Deng H (2009) Integrated assessment metrics of urban green spaces: development and application. Urban Environ Urban Ecol (1)
8. Wang T (2011) On urban landscape – the cultural context and sustainable urban development. Gansu Gaoshi Xuebao (1)
9. Xiong Y (2007) Quantitative assessment of uncertainties in coordination between the city's living environment and economic development, with Changsha city for example. J Geogr Sci (4)

Chapter 8
Green Consumption Cities

Tianpeng Gao and Yongzhen Wang

8.1 Concept and Meaning

8.1.1 Concept and Characteristics

Below is a brief introduction to relevant concepts outside of China:

Research on green consumption can be traced back to the early 1970s when some scholars began researching ecological consumption. Green consumption is viewed in many different ways around the world. The following describe some of the more influential perspectives on China. The *Green Consumer Guide* first published in the United Kingdom in 1988 defines green consumption as an effort to avoid products that represent a threat to health and the environment, require a large amount of energy, contain unnecessary waste, contain materials from endangered animals or rare plants, or have been produced with cruel treatment of animals.[1]

British scholar Ken Peattie defines green consumerism as "the purchasing and non-purchasing decisions made by consumers, based at least partly on environmental or social criteria."[2]

The 1994 Oslo Symposium on Sustainable Consumption defined sustainable consumption, or green consumption, as "the use of services and related products which respond to basic needs and bring a better quality of life while minimizing the use of natural resources and toxic materials as well as emissions of waste and

[1] Lin and Jian [1].

[2] Peattie [2].

T. Gao (✉) • Y. Wang
Lanzhou City University, Lanzhou, China
e-mail: 549297696@qq.com; wangyongzhen1974@163.com

J. Li, T. Yang (eds.), *China's Eco-city Construction*, Research Series on the Chinese Dream and China's Development Path,
DOI 10.1007/978-3-662-48153-0_8

pollutants over the life cycle of the service or product so as not to jeopardize the needs of future generations."

International environmental experts often refer to green consumption along with the five Rs that stand for "reduce, reevaluate, reuse, recycle, and rescue."[3]

Below is a brief introduction to relevant concepts in China:

The following perspectives on green consumption are particularly influential in China.

The Green Studio defines green consumption as the position that economic activities and consumption should adhere to ecological laws in order to achieve economic development and environmental protection.

Tang Xiyang summarized green consumption as the three Rs and three Es: "reduce, reuse, recycle, economical, ecological, and equitable."[4]

Liu Xiangrong writes in his book *On Ecology Civilization*[5] that green consumption involves the consumption of green products and also all consumption that causes minimal or no harm to the environment. His book defines a green product as a generic term for products that cause minimal or no harm and follow three principles. First, the production processes for these products cause minimal or no damage or pollution to the environment. Second, these products cause minimal or no damage or pollution to the environment during or after use. Third, these products are minimally contaminated if at all.

China Consumers' Association (CCA) summarized green consumption in three points. First, consumers are encouraged to choose uncontaminated or green products that contribute to human health. Second, garbage treatment minimizes environmental pollution in the consumption process. Third, consumers value environmental protection and resource and energy conservation as they pursue convenience and comfort in life to ensure sustainable consumption. Consumption meets the needs of the current generation and of future generations.[6] Green consumption can be collectively understood as an act or process of acquiring, using, consuming, or enjoying a variety of green products and services that adhere to sustainable development specifications so that humanity's needs can be met sustainably. Green consumption covers the application of sustainable development in consumption. It covers consumer behavior and offers a modern consumer philosophy or ideology consistent with sustainable development. Green consumption can also be referred to as sustainable consumption. Green consumption can be defined in a narrow or broad sense. Green consumption narrowly refers to the consumer pattern where consumers are aware that environmental degradation has affected the quality of their lives and their lifestyles and thus consciously boycott items that damage the environment to mitigate further environmental hazards. Green consumption broadly includes buying and using green products or services, boycotting environmentally unfriendly goods, and a general environmental awareness surrounding resource conservation

[3] Green Studio [3].

[4] Tang [4].

[5] Liu [5].

[6] China Consumers' Association [6].

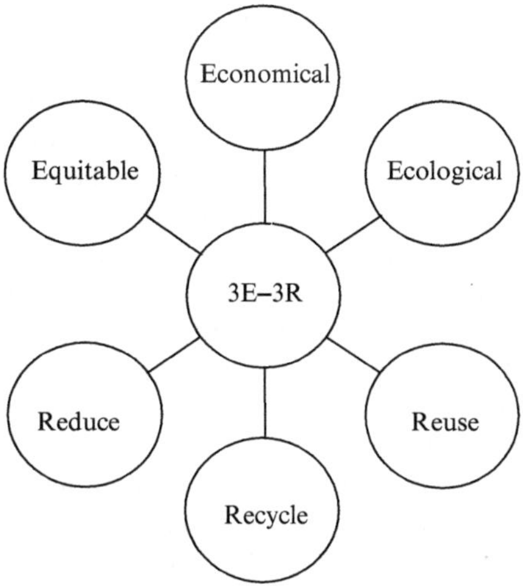

Fig. 8.1 Three Es and three Rs

and reducing pollution in the consumption process as well as theories and practices associated with green consumption and conducive to environmental protection. The broad definition of green consumption is used for the purposes of this book.

Green consumption covers all aspects of producer and consumer behaviors, including green products, material recycling, effective energy use, environmental and species protection, etc.[7] Green consumption has three characteristics. First, consumers are encouraged to choose uncontaminated or green products that contribute to human health. Second, consumers are guided toward respecting nature and incorporating health considerations by valuing environmental protection and resource and energy conservation in the pursuit of comfort in life, which encourages them to adopt sustainable consumption. Third, garbage treatment minimizes environmental pollution in the consumption process. These three characteristics match with the three Es and three Rs (Fig. 8.1), namely, economical, ecological, equitable, reduce, reuse, and recycle.[8]

8.1.2 Concept and Characteristics of Green Consumption Cities

Concept Eco-cities are an important part of a socialist harmonious society, and green consumption cities offer new opportunities in eco-city development. A strict definition of green consumption city that is generally accepted does not yet exist.

[7] Zhu [7].

[8] He and Quan [8].

This book proposes a three-part definition for a green consumption using the concept of green consumption, its background conditions, and eco-city characteristics. First, consumers are encouraged to choose uncontaminated or green products that contribute to human health to direct new urban development. Second, wastes treatment and disposal are emphasized in the consumption process as industrialization is reformulated. Third, consumers are guided toward respecting nature and considering their health by valuing environmental protection and resource and energy conservation as they pursue comfort in life, thus establishing urban sustainable consumption while working toward a harmonious society.

Green consumption has been gradually integrated across society, although significant effort from individuals and organizations remains in developing green consumption cities.

Consumers shape green consumption cities. Green consumption cities are closely tied to the public and every urban resident must promote green consumption. Whether consumers adopt green consumption or not directly determines a city's development trajectory relative to a green consumption city. The state should promote green consumption and disseminate green consumption information to educate the public about green consumption benefits. Green consumption knowledge should be popularized among consumers using institutionalized and systematic education programs. Urban residents need help enhancing their environmental awareness and promoting changes to consumer ideology regarding green consumption, thus establishing a diverse and reasonable green consumption pattern.

Businesses enable green consumption cities. Green products are produced by businesses. All businesses should adapt to consumers' desire for green products and legally profit from it. China has become a competitive buyer's market, forcing businesses to seize opportunities such as the green consumption trend and create demand for their products in order to gain a foothold in the market. They should gradually reduce the costs and prices of green consumer products by employing mass production to encourage consumers to buy the products.

The government manages and guides a green consumption city. Green consumption cities require transforming the current consumption pattern, especially impulse buying, into sustainable and rational consumption. Consumers or businesses will struggle to achieve such a transformation along because not all individuals possess the rationality and moral sense required. Government action is needed in the event of market failure and in the transformation toward green consumption. The government has an important role in green consumption. It can encourage the use of green technology and green product manufacturing through taxation, forming green industries, meeting society's needs, and fueling economic restructuring. The government can make laws to test and inspect products so as to ensure the authenticity of green products as they become popular and profitable, avoiding fake and shoddy products. It can also promote green products and consumption to encourage good consumer behavior in the public. Laws that favor green consumption allow the government to regulate consumer behavior and restrict nongreen consumption as it hinders sustainable economic development and sustainable consumption. This assures the systematic development of green consumption across society and creates a healthy, comfortable, and environmentally friendly consumption environment.

Green consumption is necessary for urban development. Changing the consumption pattern and lifestyles is necessary in developing eco-cities and for sustainable development. Agenda 21 declared in 1992 that consumption is at the core of environmental crises. The consumption pattern must be changed for the world to address environmental crises. Green consumption is necessary for developing eco-cities.

Environmental damage and the massive consumption of resources and energy are reflected in various products, such as energy-consuming buildings and home appliances, vehicles with low gas mileage, overly packaged foods, disposable containers, and bottled water, which all damage the environment and generate a lot of garbage. Nearly 3,000,000 t of garbage are generated annually in Beijing, including about 830,000 t of product packaging, of which 600,000 t are excessive. Packaging accounts for 30 % of all MSW generated annually across China.[9] Separate data shows that disposable chopsticks consume as much as 1.66 million cubic meters of timber in China per year. Areas in northern China consume 18 million tons of standard coal for heating in winter each year due to many energy-inefficient buildings, which translates into a direct economic loss of 7 billion yuan and 520,000 t of carbon dioxide emissions. Buildings across China are expected to consume the energy equivalent of 108.9 billion tons of standard coal in 2020 if these trends continue. Air conditioning in summer will consume ten times as much power as the generation capacity of the Three Gorges Hydropower Station.[10] This consumption pattern will inevitably exacerbate inefficient production conflicts with mass consumption. Consumer demand may be satisfied but various urban problems have appeared, including an enormous resource waste, worsening environmental pollution, and a sharp loss of species diversity.

Green consumption results from humans rethinking how they relate to nature. In the relationship between humanity and nature, humans act as developers, users, or destroyers of nature depending on the consumer behavior and the consumption pattern. Green consumption emphasizes moderation, avoiding or minimizing environmental damage, respecting nature, and protecting ecosystems. This consumption pattern will lead production toward a circular economy and industries toward green production, establishing a firm basis for eco-cities and supporting sustainable urban development.

Green consumption is integral to urban development. The reciprocal causation between production and consumption drives and sustains socioeconomic activities. Production provides for consumption, which promotes production in a cyclical fashion. Today's buyer's market favors consumption over production. Consumption has become pivotal to social reproduction. Consumption patterns, mixes, and ethical values significantly affect the production patterns and social institutions. Disposable items will not be produced if they are not consumed. If every consumer boycotts disposable items, no enterprise will continue to produce them, reducing resource consumption and environmental pollution and facilitating sustainable urban development.

---

[9] Huang [9].

[10] Wang [10].

The characteristics of a green consumption city include:

Healthy, Harmonious, Efficient, and Energetic A green consumption city is ecologically harmonious, has an efficient local life support system, and is capable of providing normal and stable living services, sustaining accumulated surplus production. A green consumption city will change the high consumption, one directional operation mechanism of modern cities by efficiently utilizing land, human, and other resources. Materials and energy will be circulated at multiple levels and waste will be recycled. Sectors will be well coordinated to enable high productivity and positive production relations.

A Virtuous Circle Between Production and Consumption Optimizing the consumption and industry mixes furthers green consumption development and green consumption cities. A green consumption city furthers the development, promotion, and use of green products by developing green industries and promoting green consumption, thus establishing a virtuous circle between production and consumption while fueling economic growth.

Highly Eco-friendly Green consumption city culture is highly eco-friendly, containing eco-friendly communities where humans and nature are in harmony. Residents in such cities behave in an ecologically ethical manner because they have embraced green consumption. Businesses in a green consumption city are deeply aware of environmental protection and supply green products made using eco-friendly production processes. Governmental departments use their role as urban managers to regulate and direct urban development in order to guide, control, and supervise the process.

Comprehensive Human Development Green consumption is significant for human development since it radically changes humanity's values and lifestyles. Comprehensive human development is easy in a green consumption city because residents improve themselves while using green products in a beautiful living environment.

A green consumption city has other eco-city characteristics such as sustainability, integrity, and regionality.

The green consumption city pattern in China is as follows:

China has incorporated green development into its national strategies. The theme for the Bo'ao Forum for Asia Annual Conference 2010 was "Green Recovery: Asia's Realistic Choice for Sustainable Development." The attendants discussed pressing economic issues for Asia and the rest of the world such as Asia's development pattern, cooperation between emerging economies, global economic governance, the postcrisis private sector development, and low-carbon energy sources. The then Chinese Vice President Xi Jinping attended the conference's opening session and delivered a keynote speech entitled "work together for green and sustainable development in Asia." The speech stressed that

> Asian countries should coordinate with each other for economic growth, social development, and environmental protection. China is willing to cooperate with the other Asian countries in six fields including: changing the development pattern to achieve green

> development; uplifting science and technology and promoting innovation; further opening up and achieving harmonious development; enhancing cooperation for further development; working to eradicate poverty and gradually achieving balanced development; and developing human resources and promoting comprehensive development.

The UNDP's *China Human Development Report 2002, Making a Green Development Choice* first proposed that China should adopt green development to replace the growth first development pattern that has existed for so long. The traditional pattern is considered very dangerous and leads to increasingly serious environmental problems and highly uneven development by allowing pollution and then seeking treatment, placing a particularly heavy burden on the poor. China can skip over the high resource consumption and pollution that Western countries have experienced and push directly toward green development. It can begin implementing its green development strategy before achieving higher national income levels. Human development and green development are at the core of world development in the twenty-first century. China cannot modernize under the traditional development pattern nor replicate Western successes. Dr. Hu An'gang, Director of the Center for China Studies, Tsinghua University, explains that Europe, the United States, Japan, and other countries connected themselves to global markets by actively participating in international trade as they industrialized. He clarifies that these countries and regions generally relied on war and colonization to accumulate initial capital and exploit international markets, achieving economic takeoff using industrialization and international trade. China faced highly competitive global markets when it began opening up after a century of ordeals. An export mix dominated by primary commodities put China in a highly disadvantageous position. China is also late to industrialize and remains behind industrialized countries with regard to production technology and resource development and utilization capacity. It faces enormous industrialization difficulties despite profiting from latecomer advantages. China's per capita resource possession is only a fraction of the rate found in European countries or the United States, hindering its capability to achieve modernization using high resource and energy consumption in production and in life. China cannot accumulate capital through war and plundering resources like industrialized countries did in the past. It can only advance using internal reform and development. Dr. Hu argued that

> China's development strategy is bound to differ from the traditional modernization pattern. It should adopt a non-traditional development pattern that accounts for national conditions. At the core of this pattern are a low resource consumption production system; moderate consumption in life; an economic system that enables sustained, steady economic growth and increased economic efficiency; a social system that ensures social benefits and justice; a technology application system that is continuously innovating and that is open to absorbing new technologies, processes, and methods; a more open international trade system that promotes close connections between China and other markets; as well as rational utilization of resources, pollution prevention, and preservation of ecological balance.

The environmental impact of cities is increasingly relevant to environmental efforts and future urban development. The objectives of green consumption cities include: healthy human and natural development through optimized urban functionality, full access to clean and efficient resources and energy sources, a healthy and pleasant natural environment, comfortable and reliable infrastructure, as well as a civilized and harmonious social environment. Some hold that green consumption cities should go beyond coordinating urban development with the natural environment because these cities focus on more than ecological balance and nature conservation, expanding into human health and cultural development. Green consumption cities should satisfy a number of prerequisites and proposed objectives such as: linking biological materials and cultural resources in a harmonious relationship that can support itself and preserve ecological balance; ensuring that the natural environment is viable, energy balanced, and able to produce additional energy for value creation; protecting natural resources and minimizing and recycling waste; providing humans vast natural spaces to share with a diversity of species; maintaining human health and promoting healthy lifestyles; planning and arranging elements aesthetically; ensuring beauty and rhythm in human settlements; and developing culture comprehensively, ensuring it is joyful and able to progress.

As mentioned in Chap. 2, the Chinese environmental pioneer Qu Geping proposed six criteria and nine approaches for building green consumption cities. The six criteria include: appropriate plans and spatial arrangements; reliable urban infrastructure; effective pollution prevention and healthy environmental quality; clean energy utilization; a certain percentage of landscape coverage in the city; and residents with a strong environmental awareness. The nine approaches include: developing appropriate urban plans; implementing an economic development pattern centered around a circular economy; building fully functional urban environmental infrastructure; establishing a fast, convenient, and clean urban transportation system; creating an energy system led by clean energy sources; building beautiful and eco-friendly residential areas with necessary services and high level of environmental quality; researching and developing an environmentally friendly technical support system; improving the sustainable development legal system; and raising awareness of environmental protection and resource conservation across society, promoting eco-friendly values and green consumption, and helping public figures, especially government leaders, accept sustainable development principles and meet environmental requirements in their decisions and consumption choices. These criteria and approaches determine developmental goals for green consumption cities and also offer a development strategy that stems from environmental protection and ecological perspectives, thus providing important guidance for green consumption city development in China.

8.2 Assessing Green Consumption Cities

The assessment metrics for green consumption cities form a collective unit that consists of a series of interrelated indicators that quantitatively reflect a city's characteristics with respect to sustainable consumption from different points of view. Humanity's consumption activities are not isolated but rather closely interrelated with socioeconomic activities and are continuously interacting with the natural environment. Human beings need resources and energy from the environmental system in order to survive and reproduce. They also need to meet their own needs through production activities. Human beings generate waste as part of consumption, lowering the quality of resources and the environment. These changes to environmental conditions and structures affect humanity's consumption. Evaluation of green consumption in an eco-city must consider more than consumption statistics and trends. Various factors that lead to non-sustainable consumption must be incorporated and the status and results of human efforts to restrict detrimental behavior must be examined to properly determine the extent of sustainable consumption in different countries and regions.

8.2.1 Principles for Developing Assessment Metrics

The assessment metrics for green consumption cities should reflect green consumption in the city and evaluate the extent of urban sustainable consumption. The following principles are proposed for assessment metrics to ensure green consumption characteristics are expressed properly and green consumption in an eco-city is assessed accurately.

Systematic Humans' behavior in economic, social, and environmental systems should be studied in preparation for green consumption city development. Creating green consumption cities aims to coordinate and balance humans and these three systems, so that humans will be in harmony with nature on a sustainable basis. The assessment metrics must therefore be systematic. Consumption by urban residents should first be observed to determine what economic, social, and environmental changes have occurred as a result of this consumption and what positive and negative effects these changes have on humans. Efforts by urban residents toward sustainable consumption should also be noted, as well as how these efforts affect the virtuous cycle of human beings and promote the coordination and development of human beings and the economic, social, and environmental systems and human beings' rationality while conducting various activities. These observations can guide human beings toward choices for future socioeconomic activities. It is possible to comprehensively describe and evaluate the status and extent of urban development from an ecological perspective by observing indicators of human consumption, socioeconomic development, environmental pressures from irrational human behavior, and harmony with nature in the economic, social, and environmental systems.

Coordination Green consumption is a subsystem of sustainable development and must be coordinated with the sustainable development system and the other subsystems, especially ecosystems. The assessment metrics should restrict or disallow consumer behavior that causes resource waste and environmental and ecological damage and that is incompatible with economic growth and social development, thus guiding consumption toward balanced and harmonious development with production and natural ecosystems.

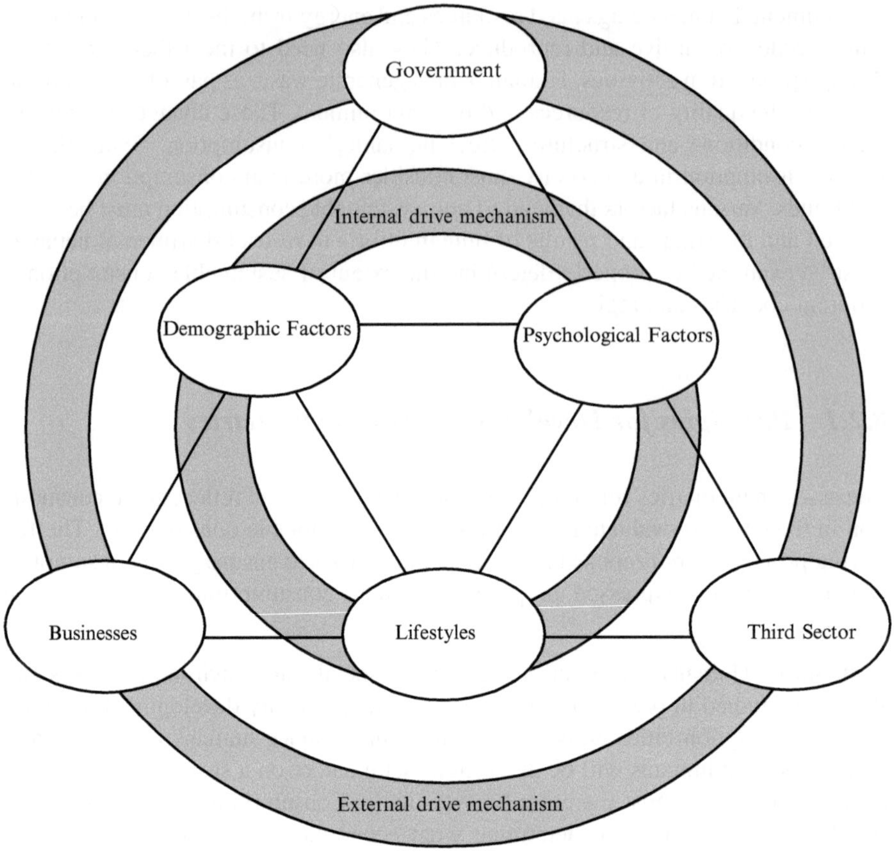

Fig. 8.2 Development mechanism for green consumption cities

Equity Intergenerational and intragenerational equity in consumption is emphasized in current development. Equity lays a firm social foundation for sustainable development. The assessment metrics should reflect social equity and encourage people to consider their personal interests in the context of others' interests and of society's interests. People should also consider the interests of future generations.

Comprehensive Urban development's goal is to improve the quality of life for all residents and to achieve comprehensive human development. The quality of life covers material and spiritual dimensions, the consumption quantity and mix, the consumption of consumer goods, and consumers' physical and psychological well-being afterward. The metrics should reflect urban development from multiple perspectives. Comprehensive assessment does not mean that the metrics present all characteristics of urban development, which is neither possible nor necessary. The focus is placed on indicators that reflect the primary characteristics of green consumption city development (Fig. 8.2).

Green consumption cities' internal drive mechanism refers to the process in which green consumers' qualities impact their purchasing behavior. The consumer qualities fall under demography, psychology, and lifestyle. The external drive mechanism refers to the process in which other factors affect green consumers' green purchasing. These driving factors include the government, businesses, and the third sector. Social, cultural, and consumer environments created by the government are important determinants of green consumers' purchasing behavior. The government is able to create green social and cultural environments for consumers by employing publicity and education. Social and cultural environments have a broad and profound impact on consumer behavior. The degree to which a society and its culture are green directly affects environmental awareness and hence consumer behavior in all social groups. Consumers in particularly environmentally friendly social and cultural environments are more likely to become green consumers. Policies conducive to a green consumption pattern will further improve the consumer environment, which increases consumer demand for green products and services and alters consumer mentality, thus enhancing their environmental awareness and promoting green consumer behavior.

The third sector includes environmental organizations and the mass media, which also have an important impact on green consumption through publicity and education. The third sector can provide consumers with environmental knowledge to enhance their environmental awareness. The third sector can also provide corporate decision makers with environmental knowledge to redirect their priorities. Educating different individuals in a fashion tailored to their needs allows the third sector to enhance society's environmental awareness and create highly green social and cultural environments for consumers, promoting the green consumption pattern.

8.2.2 Assessment Metrics for Green Consumption Cities

The green consumption city development mechanism discussed was used to establish the assessment metrics for green consumption cities that adhere to the stated principles (see Table 8.1).

This metrics set is composed of six sub-modules, aggregate and structural economics, human development, consumer environment, green urban infrastructure, consumer recognition and behavior, and green consumption regulation and support, and includes a total of 36 indicators covering the government, businesses, the third sector, and urban residents. These indicators comprehensively evaluate green consumption cities by addressing their primary criteria.

Table 8.1 Assessment metrics for green consumption cities

Type	Indicator
Aggregate and structural economic indicators	(i) GDP per capita (ii) Investment per unit area (iii) The percentage of high-tech products in GDP (iv) Per capita consumer goods retail sales (v) The percentage of the tertiary sector's added value in GDP (vi) Outstanding loans to financial institutions
Human development indicators	(vii) Per capita disposable income (viii) Per capita floor space (ix) Per capita power consumption (x) The number of undergraduates per 1000 persons (xi) The number of ward beds per 1000 persons (xii) The number of criminal cases per 1000 persons (xiii) The per capita area of green space
Consumer environment indicators	(xvi) The percentage of nonrenewable resources recovered from consumption in daily life (xv) The percentage of renewable energy sources out of total energy consumption (xvi) The number of years for which nonrenewable resources will be available (xvii) The average level of ambient noise and the percentage of eliminated ambient noise (xviii) The composite urban air pollution index (xix) The utilization of drought-tolerant and native plants (xx) The preservation of conditions and environmental factors for the healthy growth of plants (xxi) The protection of park and shading plants and of large diameter trees
Green urban infrastructure indicators	(xxii) Facilities for pedestrians, public transportation, and cycling (xxiii) The application of low-reflecting materials and of landscaping on rooftops (xxiv) The application of permeable materials on the ground (xxv) The application of environmentally friendly and energy-efficient building materials (xxvi) Energy efficient, low-pollution building design
Consumer recognition and behavior indicators	(xxvii) Green consumption recognition (xxviii) The number of businesses and products eligible for green labeling (xxix) The number of green markets at the national level (xxx) The number of green markets at the provincial level (xxxi) The distance raw materials travel for extraction, production, and transportation (xxxii) The number of motor vehicle passengers (xxxiii) The use of pollution sources (e.g., refrigerators containing CFCs)
Green consumption regulation and support indicators	(xxxiv) Green consumption regulators (xxxv) Regulations regarding green consumption (xxxvi) Incentives for traveling by foot, by public transportation, or by cycling

8.3 Strategies and Measures for Building Green Consumption Cities

8.3.1 Strategies

Transforming the consumption pattern to green consumption will change the production pattern and ensure green products are used, green industries developed, and green homes built in the city. Advocating green consumption can reduce waste at the source and effectively promote sustainable urban development and eco-city development.

Establishing Green Consumption Among All People and Encouraging Eco-City Building Participation Green consumption has gradually gained public acceptance and quality of life has risen and environmental awareness has spread. Green consumption nevertheless remains far from the dominant lifestyle, requiring further efforts from the entire society. People's adoption of green consumption directly determines its market penetration. The state should publicize green consumption and disseminate information and knowledge about green consumption to clearly express its benefits. The public's environmental awareness should be enhanced through institutionalized and systematic education programs, thus shifting consumer ideology toward green consumption and establishing a reasonable and diverse green consumption pattern. Unlike the traditional consumption pattern, in which consumers only care about personal benefits and spare little concern for public and environmental benefits, green consumption combines consumer benefits and environmental protection to ensure sustainable human survival. Green consumers do not approve of sacrificing the environment for consumer benefits and hence resist purchasing and consuming goods that pollute the environment. A green consumption lifestyle with public approval promotes public participation in eco-city development and will be very beneficial for harmonious development between humans and nature and among humans.

Adjusting Industry and Consumption Mixes Adjusting the industry mix is critical to building eco-cities and requires environmental protection that complies with high standards. The traditional industry mix suffers from high energy consumption and severe pollution and so must change. The government should adjust the energy and industry mixes, promote green energy sources, and encourage the development of environmentally friendly high-tech industries in the tertiary sector to create a green urban environment. Adjusting the industry mix will also improve the consumption mix and pattern. A green consumption pattern will lead to a green lifestyle with an improved quality of life and an environmentally friendly city.

Promoting Green Consumer Behavior to Protect the Urban Environment The urban environment is closely connected to human consumption. Green consumption can avoid or mitigate environment pollution caused by improper urban planning, an irrational industry mix, an unbalanced urban layout, inefficient urban management, and poor public facilities and equipment. Green consumption insists on low energy and pollution production processes, green product development, and utilization of green packaging and green communication. Green consumption emphasizes rational consumption where consumers consider personal preference as well as social and ecological consequences since they attach equal weights to personal and public benefits, which is not present in traditional consumption patterns. Green consumption effectively reduces consumption's negative impact on the urban environment and is conducive to eco-city construction. Eco-cities require a reciprocal green consumption pattern present in consumer behavior.

Using Resources to Promote Sustainable Eco-City Development MSW is increasing rapidly due to spreading urbanization. Many recyclable materials end up in MSW although they can be made into new products with a small addition of mineral resources. The mix creates a dynamic balance between scraped and recycled materials. Green consumption will produce vital green effects in eco-city development. Green consumption goes beyond green products to include material recycling, effective energy use, and environmental and species protection. It emphasizes the "Save and Reduce" (SR) principle and avoids excessive, irrational, and wasteful consumption and successfully recycles resources, effectively protecting the city's resources and environment.

Improving the Consumption Mix and Quality of Life to Create a Harmonious Eco-City Consumption is a crucial component of social reproduction that is dictated by production but also influences it. Green consumption transforms the high resource and energy traditional consumption pattern into a new consumption pattern that uses resources and energy efficiently with a goal to further improve people's lives. Improvements to the consumption mix benefit a green living environment and a harmonious society, promoting sustainable urban development. Green consumption is a prerequisite for eco-city development.[11]

8.3.2 Measures

Green consumption is new to China requiring inventive green consumption cities concepts. This book proposes some measures for consumers, the government, and businesses to help promote green consumption and develop Chinese cities into green consumption cities.

Consumers Should Embrace Green Consumption and Purchase Environmentally Friendly Products Popularizing green consumption requires publicity and education to change deep-rooted traditional consumer behavior and establish an atmosphere conducive to green consumption, according to a green consumption survey in China. Communities, businesses, and other organizations should regularly offer activities to disseminate green consumption knowledge and advocate

[11] Mingzhi [11].

green consumption. Consumers require guidance toward a rational consumption pattern and green consumption. Corresponding measures may include: publicizing resource conservation and pollution reduction, encouraging green consumption and purchasing and reusing materials, rescuing species to preserve nature, and avoiding goods that endanger consumer health; that require a large resource input for production, use, or disposal; that cause unnecessary consumption due to excessive packaging or a short lifecycle; that use materials from rare animals or natural resources; that engage in animal cruelty during production; and that have an adverse impact on other developing countries. Water-conserving sanitary products, energy-efficient appliances, phosphate-free detergents, and cloth shopping bags should all be promoted to further these efforts, as should boycotts of overly packaged goods and water reuse. Community environmental efforts such as garbage and waste sorting and recycling should be encouraged. Establishing green consumption fosters green consumerism and a market for green products, thereby promoting eco-city development.

Green Business Operations British economist Alfred Marshall theorized that all needs are eventually mediated by consumer demand. Businesses combine business, consumer, social, and environmental interests, implement green marketing strategies, and develop green corporate values to establish a green corporate image. Green operations require companies to conduct production and management in a way that maintains ecological balance, improves environmental quality, and satisfies consumer needs with green products. Companies should focus on cleaner production and green marketing. Cleaner production of green products should create little or no environmental pollution and resource waste. These products should cause no threats to human health or the environment during and after use. Companies should conduct market research, product pricing, sales channel development, and promotional activities with consideration of ecological balance and environmental protection. Companies should communicate green corporate information to the public through public relation activities as they provide services to meet consumer needs, thus increasing influence in the market and establishing a corporate image as a green product provider. Companies should also work to receive environmental labeling and green product certification, especially under ISO14020.

The Government Should Enhance Managerial Functions Green industries and green consumption development require encouragement and support. The government should establish developmental objectives and priorities for green industries and green consumption in the city, phase in relevant plans, and position green industries as a new growth engine. The government should draft policies, laws, and regulations to support green consumer market growth. For example, China has promulgated more than 20 specific environmental laws such as the *Environmental Protection Law*, the *Marine Environment Protection Law,* and the *Law on Prevention and Control of Water Pollution,* as well as administrative regulations such as the *Regulations on Prevention and Control of Environmental Noise Pollution*. Plastic bag bans, the *Regulations on Energy Conservation in Civil Buildings,* and the *Regulations on Energy Conservation in the Public Sector* are further examples of related government action. These laws and regulations are far from satisfying the requirements for green consumption and beg further improvements. Inspection and supervision must be enhanced and producer and consumer behavior regulated. China's green production technologies, production environment standards, and product quality standards require immediate development to meet internationally accepted standards and facilitate green product certification. Leading green product brands should be promoted and green product patents protected. The green government procurement system should be effectively implemented, and market mechanisms should guide producer and consumer behavior and raise environmental awareness across society, achieving sustainable urban development. The government is also responsible for ensuring that efforts to do rely heavily on the market. An important market mechanism measure is modifying energy and resource pricing reform to properly reflect market supply and demand. The government must use past medical and educational reforms as lessons to learn from. The government remains obligated to provide the public with services connected to vital public interest across society.

Improve Fiscal and Tax Policies The government needs to restructure its fiscal and tax policies. It should offer higher tax credits to companies that use renewable resources, manufacture green products, and recycle used materials. It should provide policy incentives for investment in energy-efficient and resource-conserving technologies and equipment as well as offer comprehensive rebates for export of circular economy products. The government also needs to further increase the resource tax. It is estimated that a 30 % increase in the resource tax would have little impact on GDP, no more than 0.3 percentage points, but would promote a marked reduction in resource consumption, technological improvements, environmental protection, and green consumption. Export tariffs should be increased for products that use a lot of energy and resource or cause severe pollution and for primary commodities, and the value-added tax (VAT) should be transformed to a consumption tax.

Enhance Green Design Green consumption is built upon green design. Green design is a priority for the manufacturing industry and for the building and construction industry. The construction industry is one of the highest resource and energy consumers; almost half of China's steel production is consumed in this industry. The power and resource consumption of a building is determined after it is designed, and current green design capacity in China is very low. Statistics indicate that two-thirds of China's new oil demand in the past year comes from the transportation sector, where automobile oil consumption accounts for over one-third. Enhancing automotive energy-efficient technology design and development will significantly restrict excessive oil consumption growth.

Expand the Waste Recycling Sector Waste recycling in China relies on a large number of rural-urban migrants to gather and process materials. The waste recycling sector is not without its achievements but faces problems including the physical health of rural-urban migrants, secondary pollution, and inefficient processing technologies. Developing the waste recycling sector and the larger circular economy is intended to solve environmental problems, and increased environmental pollution runs contrary to this intention. The state should provide technical standards to prevent secondary pollution in recycling development. Japan, Germany, and other countries have noticed that a developed waste recycling sector provides green consumption support and is an excellent job creator because of the labor-intensive nature of the recycling process. The state should draft policies and unified arrangements covering land use, tax relief, government policy, product and technology approval, and R&D investment in a manner similar to the one it adopts in other industries to effectively grow the waste recycling sector.

8.4 Dalian: A Green Consumption City

China was late to begin developing green consumption but is gradually improving. In the 9th Five-Year Plan period, China formulated and began to implement China's Cross-Century Green Initiative. The then SEPA and five other government authorities launched programs that established green channels, cultivated a green market, and promoted green consumption. The 10th Five-Year Plan also proposed strategically elevating environmental development and protection to achieve sustainable development. In 2001, the CCA announced their theme of the year to be green consumption and committed to popularizing it. This theme will remain pertinent throughout the twenty-first century. China's 11th Five-Year Plan explicitly expresses a resource-conserving and environmentally friendly society as a developmental goal. The report presented by Hu Jintao at the 17th National CPC Congress elaborated on eco-civilization development, emphasizing the eco-friendly values and efforts toward an eco-civilization across society. This report is extremely significant for promoting eco-civilization development theories and practices in China. The outline of the 12th Five-Year Plan proposed promoting green consumption, green lifestyles, and consumption patterns compatible with China's national conditions. Advocating a green consumption pattern is key to implementing the Scientific Outlook on Development and developing toward an eco-civilization. Dalian's development has been dependent on green growth. The municipal government focuses on developing low-carbon, green industries to advance green economic growth. It encourages citizens to adopt low-carbon lifestyles and protect the environment to advance green social growth. The Blue Sea and Sky program aims to leave the city greener and more beautiful and livable as it develops. Dalian is an old industrial city in Northeast China that suffered from severe pollution and which had traditional industries that required renewal and rejuvenation. The municipal government chose green growth and has attempted to eliminate green urban development bottlenecks. Dalian popularized the ideas of "green," "low-carbon," and "new energy sources" as the economic development pattern was being rapidly transformed and the city was developing toward a green future.[12]

[12] "The green rise of Dalian," the People's Daily, March 7, 2013.

Developing a green and low-carbon economy allows Dalian to implement the Scientific Outlook on Development in its relation to sustainable socioeconomic development. Dalian has achieved advanced and coordinated socioeconomic and environmental development. This stage is important for Dalian to evolve into a green consumption city. The municipal government has been orchestrating efforts to optimize the local industry mix; to increase the new energy source shares, including nuclear and wind power, in primary energy consumption; and to promote green technologies such as LEDs and photovoltaic power generation, thereby reducing overall energy consumption and pollution. Concepts such as "green," "low-carbon," and "green consumption" have been included in the local socioeconomic development program and have shaped the local policy framework in response to climate change, disaster reduction, and sustainable development. Comprehensive arrangements of new energy sources, green industries, energy-efficient buildings, sustainable transportation, and consumer behavior are undertaken to form a framework for low-carbon, green economy development and Dalian's green consumption urban development.[13,14] Dalian's has had a number of experiences in green consumption city construction:

Improve the Industry Mix Measures for this purpose include: optimizing the energy conservation management system in industrial enterprises; promoting EPC to improve energy management and efficiency; developing high-tech industries and service industries that consume low amounts of energy and produce high added value; and attracting worldwide research organizations to open laboratories and innovation centers in Dalian, thus promoting green technology R&D and application and making Dalian a green technology R&D and innovation hub.

Optimize the Energy Mix Efforts have accelerated clean energy and emissions reducing technology R&D and application, reduced Dalian's reliance on fossil fuels for economic development, and promoted the high carbon energy chain transformation to low carbon, making Dalian a new energy sources hub in Northeast China.

Improve the Multidimensional Public Transportation System Measures for this purpose include: optimizing the public transportation network plan; prioritizing rail transit, subway, and bus rapid transit (BRT) development; promoting new energy vehicles; balancing road resources between public transportation and motor vehicles; and testing a motor vehicle congestion fee. These measures are intended to form a multidimensional network for public transportation operation and management offering convenient and efficient transfers.

Encourage Green Building Development Measures include: promoting solar, LED, and other green technology application; encouraging construction of eco-friendly apartment complexes with efficient water use, land use, energy consump-

[13] Dalian Eco-friendly Development Plan, [R].2011.

[14] The Dalian 12th Five-Year Plan of Socioeconomic Development, [R].2010.

tion, and sewage treatment; modifying existing public buildings to conserve energy; and introducing the green building labeling and evaluation system.

Encourage the Public to Adopt Green Lifestyles The municipal government encourages the public to use public transportation and subsidizes energy-efficient consumer goods. It encourages public participation in afforestation to grow carbon sinks and reduce the carbon footprint by increasing forest coverage.

Improve the Carbon Trading Mechanism and Green Financial Services The market is allowed a fundamental role in promoting green economic development. An exchange for the right to discharge or emit pollutants has been opened in Dalian. Carbon and weather derivatives trading is accomplished through local futures exchanges to supplement professional services, market infrastructure, and transaction settlement systems. This practice seeks to align carbon prices with resource scarcity and pollution prevention and control costs. Effective carbon asset management have been sought with efforts like the Clean Development Mechanism (CDM), and carbon benefits have been offered to replace carbon costs in economic development, making Dalian a regional center for pollution and carbon emissions permit trading. Funds are allocated to favor green industries by allowing banks to allocate credit, insurers to provide transfer and risk avoidance services, and institutional investors to undertake fiduciary responsibility for environmental governance.

Enhance International Cooperation on the Green Economy The municipal government of Dalian has long cooperated on green economy development with international organizations such as the United Nations, the World Bank, the Asian Development Bank, developed countries, and other cities. Measures have attracted international funding and technology transfer, established bilateral and multilateral cooperation promotion centers, and strengthened local green development capacity. Green is the new emblem of a romantic and stylish Dalian, symbolizing life and vitality. Dalian has profited from opportunities offered by the central government to revitalize old industrial cities in Northeast China and develop the Liaoning coastal economic zone to pursue green growth and green urban development. A green consumption city covers many ideas and all aspects of urban development. It aggressively develops low-carbon industries that produce little or no pollution, such as modern services and new energy sources. The government encourages citizens to adopt low-carbon lifestyles and protect the environment. The Blue Sea and Sky program is promoted to make the city greener and more beautiful and livable. Urban landscaping, urban economy, and citizens' lifestyles should all work to "go green." Green consumption city construction can become more efficient only through public participation by adopting green lifestyles in all aspects, such as home improvement and garbage disposal. Signs indicate that Dalian is undergoing a green revolution. This revolution will produce a profound impact on the city's economic development as well as the daily lives of ordinary citizens (Table 8.2).

Table 8.2 Top 50 Chinese cities by overall index of green consumption city, 2008–2011

2008		2009		2010		2011	
City	Ranking	City	Ranking	City	Ranking	City	Ranking
Shenzhen	1	Shanghai	1	Shenzhen	1	Shenzhen	1
Shanghai	2	Beijing	2	Shanghai	2	Guangzhou	2
Nanjing	3	Shenzhen	3	Beijing	3	Xi'an	3
Beijing	4	Nanjing	4	Guangzhou	4	Shanghai	4
Guangzhou	5	Guangzhou	5	Nanjing	5	Beijing	5
Xiamen	6	Hangzhou	6	Hangzhou	6	Nanjing	6
Dongguan	7	Xiamen	7	Dalian	7	Zhuhai	7
Wuxi	8	Tai'an	8	Fuzhou	8	Hangzhou	8
Haikou	9	Fuzhou	9	Chongqing	9	Xiamen	9
Jinan	10	Dongguan	10	Tai'an	10	Dongguan	10
Suzhou	11	Dalian	11	Xiamen	11	Dalian	11
Yantai	12	Tianjin	12	Suzhou	12	Wuhan	12
Shenyang	13	Shenyang	13	Weihai	13	Tianjin	13
Dalian	14	Zhuhai	14	Shenyang	14	Ningbo	14
Wuhan	15	Dongying	15	Yantai	15	Jinan	15
Hohhot	16	Weihai	16	Changchun	16	Shenyang	16
Changsha	17	Suzhou	17	Chengdu	17	Changsha	17
Qingdao	18	Nanning	18	Nanning	18	Nanchang	18
Changchun	19	Chongqing	19	Wuxi	19	Yantai	19
Chengdu	20	Taiyuan	20	Quanzhou	20	Changchun	20
Nanning	21	Wuxi	21	Dongying	21	Baotou	21
Chongqing	22	Jinan	22	Baoding	22	Hefei	22
Ma'anshan	23	Changchun	23	Shaoxing	23	Wuxi	23
Dezhou	24	Wuhan	24	Dongguan	24	Changzhou	24
Nanchang	25	Yantai	25	Tianjin	25	Jilin	25
Weihai	26	Chengdu	26	Ningbo	26	Yinchuan	26
Ningbo	27	Hohhot	27	Qingdao	27	Chengdu	27
Tai'an	28	Ningbo	28	Zhuhai	28	Jiaxing	28
Tianjin	29	Dezhou	29	Jiaxing	29	Harbin	29
Taiyuan	30	Qingdao	30	Hohhot	30	Fuzhou	30
Jiaxing	31	Shijiazhuang	31	Zhongshan	31	Hohhot	31
Zhengzhou	32	Baoding	32	Changzhou	32	Zhengzhou	32
Fuzhou	33	Shaoxing	33	Jinan	33	Haikou	33
Yinchuan	34	Xi'an	34	Yinchuan	34	Suzhou	34
Hangzhou	35	Changsha	35	Wuhan	35	Zhenjiang	35
Urumqi	36	Haikou	36	Daqing	36	Chongqing	36
Hefei	37	Guiyang	37	Changsha	37	Dongying	37
Daqing	38	Hefei	38	Zhengzhou	38	Weihai	38
Dongying	39	Quanzhou	39	Harbin	39	Nanning	39
Baoding	40	Jiaxing	40	Zhenjiang	40	Daqing	40
Xi'an	41	Nanchang	41	Hefei	41	Urumqi	41

(continued)

Table 8.2 (continued)

2008		2009		2010		2011	
City	Ranking	City	Ranking	City	Ranking	City	Ranking
Zhuhai	42	Yinchuan	42	Kunming	42	Tai'an	42
Shaoxing	43	Changzhou	43	Xi'an	43	Huzhou	43
Harbin	44	Ma'anshan	44	Baotou	44	Shaoxing	44
Zhongshan	45	Luoyang	45	Jilin	45	Taiyuan	45
Guiyang	46	Jining	46	Anshan	46	Zhongshan	46
Quanzhou	47	Linyi	47	Jining	47	Quanzhou	47
Jining	48	Qinhuangdao	48	Shijiazhuang	48	Guiyang	48
Xiangtan	49	Xuzhou	49	Huzhou	49	Shijiazhuang	49
Anshan	50	Guilin	50	Nanchang	50	Baoding	50

References

1. Lin B, Jian X (2000) Consumer economics dictionary. Economic Science Press, Beijing
2. Peattie K (1993) Green marketing, Tomisaburo Mikami as translation supervisor. Doyu-kan, Beijing, p 129
3. Green Studio (1998) Green consumption. Beijing Ethnic Publishing House, Beijing, p 23
4. Tang X (1993) A green world tour. Lijiang Publishing House, Guilin
5. Liu X (1999) On ecology civilization. Hunan Education Publishing House, Changsha
6. China Consumers' Association (2001) China Consumers' Association outline of publicity in the year of green consumption. China Industry and Commerce Press, Beijing
7. Zhu J (2001) On green consumption. Lingnan J (4)
8. He W, Quan X (2001) On obstacles to green consumption. Consumer Econ (2)
9. Huang J (2001) Packaging waste needs to be legally limited promptly. Leg Daily, May 23, 2001
10. Wang H (2007) Energy conservation measures should start from sources. Outlook Wkly (13)
11. Mingzhi Y (2008) Green consumption: where eco-city building starts. China Bus (6)

Bibliography

1. Beijing Normal University, Southwestern University of Finance and Economics and China Economic Monitoring & Analysis Center, the National Bureau of Statistics of China (2011) China green development index report 2011 – regional comparison. Normal University Press, Beijing
2. International Institute for Urban Development (2003) Blue book of Chinese cities. China Modern Economic Publishing House, Beijing
3. Brown LR (2002) Eco-economy: building an economy for the Earth. Oriental Press, Beijing
4. Fang Y et al (2011) SST theoretical study of eco-civilization in China. China Zhigong Publishing House, Beijing
5. Gao D (2011) Eco-civilization and sustainable development. China Zhigong Publishing House, Beijing
6. Hailing G (2012), Theoretical and empirical research on low-carbon eco-city development. Economic Science Press, Beijing
7. IEA (2004) World energy outlook – 2002. China Petrochemical Press, Beijing
8. Guo Q (2005) Resource-efficient society. China Modern Economic Publishing House, Beijing
9. Yutang H (2011) On the building of an eco-friendly industry chain. China Zhigong Publishing House, Beijing
10. Hong F (2011) Eco-civilization and environmental governance model innovation in China. China Zhigong Publishing House, Beijing
11. The Department of Nature and Ecology Conservation, The Ministry of Environmental Protection (2008) Eco-province development: theories and practices. China Environmental Science Press, Beijing
12. Huang G, Chen Y (2002) Ecocity: theory and design approach. Science Press, Beijing
13. Li B (1999) Ecology. Higher Education Press, Beijing
14. Li X, Li W (2006) Environmentally friendly cities. China Environmental Science Press, Beijing
15. The Chinese Preparatory Committee for the United Nations Conference on Sustainable Development (2009) The People's Republic of China report on sustainable development strategy. Science Press, Beijing
16. Peizhe L, Pan J et al (2001) Theory of sustainable development and China's Agenda 21. Meteorological Press, Beijing
17. Rubai M, Zhijun F (2003) On circular economy. Economic Science Press, Beijing

J. Li, T. Yang (eds.), *China's Eco-city Construction*, Research Series on the Chinese Dream and China's Development Path,
DOI 10.1007/978-3-662-48153-0

18. Pepper D (2005) Eco-socialism: from deep ecology to social justice. Shandong University Press, Jinan, Shandong
19. Bozhang Q (2008) Energy conservation and emission reduction: the only way to sustainable development. Science Press, Beijing
20. Sun J et al (2006) Sustainable urban development. China Renmin University Press, Beijing
21. Sun G (2005) The new paradigm of circular economy. Tsinghua University Press, Beijing
22. Tian Y (2010) Primary research on eco-city planning formulation. Chinese Construction Industry Press, Beijing
23. Wang R (1990) Urban ecology. Science Press, Beijing
24. Wang Y (2010) Eco-city Building: Theoretical and empirical research. China Zhigong Publishing House, Beijing
25. Wang D (2007) Outlook on China's energy technology development in the 21st century. Tsinghua University Press, Beijing
26. Wang J (2011) Environmental Law Enforcement over the past three decades: did we make it? (Blue Book of Environmental Law Enforcement in China), 1979–2010. Beijing University Press, Beijing
27. Wei Y (2006) China energy report (2006): strategy and policy research. Science Press, Beijing
28. Wu J (2006) The concept of scientific development and China's circular economy strategy. Xinhua Publishing House, Beijing
29. Chuanjun W (1998) The human-nature relationship and economic layout. Academy Press, Beijing
30. Renjian W (2000) Eco-city building: principles and approaches. Fudan University Press, Shanghai
31. Xiao D et al (2010) Landscape ecology. Science Press, Beijing
32. Chun X (2001) Sustainable development and eco-civilization. Beijing Publishing House, Beijing
33. Jacquet P et al (2010) Villes: Changer de Trajectoire. Social Sciences Academic Press (China), Beijing
34. Geng Y et al (2010) Annual report on China's provincial eco-civilization index (ECI 2010). Social Sciences Academic Press (China), Beijing
35. Yang T, Gao Y (2007) Ecological turn of modern civilization. Chongqing Publishing House, Chongqing
36. Yang Z et al (2004) Urban planning for sustainable environmental development. Science Press, Beijing
37. Zhang M (2011) On eco-civilization and its contemporary value. China Zhigong Publishing House, Beijing
38. Zhang H et al (2008) Urban positioning theory – sustainable development strategy from an urban sociology perspective. Southeast University Press, Nanjing, Jiangsu
39. Zhao Y (2008) A human-nature view in contemporary China. Northeast Normal University Press, Shenyang, Liaoning
40. Zhu T (2007) General principles of China's sustainable development (Volume 10): China's environmental protection and sustainable development. Science Press, Beijing

MIX
Papier aus verantwortungsvollen Quellen
Paper from responsible sources
FSC® C105338

If you have any concerns about our products,
you can contact us on
ProductSafety@springernature.com

In case Publisher is established outside the EU,
the EU authorized representative is:
Springer Nature Customer Service Center GmbH
Europaplatz 3, 69115 Heidelberg, Germany

Printed by Libri Plureos GmbH
in Hamburg, Germany